Real-Time Image and Video Processing: From Research to Reality

Real-Time Image and Video Processing: From Research to Reality

Nasser Kehtarnavaz and Mark Gamadia

978-3-031-01112-2   paper   Kehtarnavaz/Gamadia   Real-Time Image and Video Processing
978-3-031-02240-1   ebook   Kehtarnavaz/Gamadia   Real-Time Image and Video Processing

DOI 10.1007/978-3-031-02240-1

A Publication in the Springer series
SYNTHESIS LECTURES ON INFORMATION SECURITY, PRIVACY, AND TRUST
Lecture #5

First Edition
10 9 8 7 6 5 4 3 2 1

Real-Time Image and Video Processing: From Research to Reality

Nasser Kehtarnavaz and Mark Gamadia
University of Texas at Dallas, USA

SYNTHESIS LECTURES ON IMAGE, VIDEO & MULTIMEDIA PROCESSING #5

ABSTRACT

This book presents an overview of the guidelines and strategies for transitioning an image or video processing algorithm from a research environment into a real-time constrained environment. Such guidelines and strategies are scattered in the literature of various disciplines including image processing, computer engineering, and software engineering, and thus have not previously appeared in one place. By bringing these strategies into one place, the book is intended to serve the greater community of researchers, practicing engineers, industrial professionals, who are interested in taking an image or video processing algorithm from a research environment to an actual real-time implementation on a resource constrained hardware platform. These strategies consist of algorithm simplifications, hardware architectures, and software methods. Throughout the book, carefully selected representative examples from the literature are presented to illustrate the discussed concepts. After reading the book, the readers are exposed to a wide variety of techniques and tools, which they can then employ for designing a real-time image or video processing system of interest.

KEYWORDS

Real-time image and video processing, Real-time implementation strategies, Algorithmic simplifications for real-time image and video processing, Hardware platforms for real-time image and video processing, Software methods for real-time image and video processing

Contents

Preface

The relentless progression of Moore's Law coupled with the establishment of international standards for digital multimedia has served as the catalyst behind the ubiquitous dissemination of digital information in our everyday lives in the form of digital audio, digital images, and more recently, digital video. Nowadays, entire music libraries can be stored on portable MP3 players, allowing listening to favorite songs wherever one goes. Digital cameras and camera-equipped cell phones are enabling easy capturing, storing, and sharing valuable moments through digital images and video. Set-top boxes are being used to pause, record, and stream live television signal over broadband networks to different locations, while smart camera systems are providing peace of mind through intelligent scene surveillance. Of course, all of these innovative multimedia products would not have materialized without efficient, optimized implementations of practical signal and image processing algorithms on embedded platforms, where constraints are placed not only on system size, cost, and power consumption, but also on the interval of time in which processed information must be made available. While digital audio processing presents its own implementation difficulties, the processing of digital images and video is challenging primarily due to the fact that vast amounts of data must be processed on platforms having limited computational resources, memory, and power consumption. Another challenge is that the algorithms for processing digital images and video are developed and prototyped on desktop PCs or workstations, which are considered to be, in contrast to portable embedded devices, resource unlimited platforms. Adding to this the fact that the vast majority of algorithms developed to process digital images and video are quite computationally intensive, one requires to resort to specialized processors, judicious trade-off decisions to reach an accepted solution, or even abandoning a complex algorithm for a simpler, less computationally complex algorithm. Noting that there are many competing hardware platforms with their own advantages and disadvantages, it is rather difficult to navigate the road from research to reality without some guidelines. *Real-Time Image and Video Processing: From Research to Reality* is intended to provide such guidelines and help bridge the gap between the theory and the practice of image and video processing by providing a broad overview of proven algorithmic, hardware, software tools and strategies. This book is intended to serve the greater community of researchers, practicing engineers, and industrial professionals who deal with designing image and video processing systems and are asked to satisfy strict system design constraints on performance, cost, and power consumption.

CHAPTER 1

Real-Time Image and Video Processing Concepts

1.1   INTRODUCTION

The multidisciplinary field of real-time image and video processing has experienced a tremendous growth over the past decade, as evidenced by a large number of real-time related articles that have appeared in various journals, conference proceedings, and books. Our goal by writing this book has been to compile in one place the guidelines one needs to know in order to take an algorithm from a research environment into an actual real-time constrained implementation.

Real-time image and video processing has long played a key role in industrial inspection systems and will continue to do so while its domain is being expanded into multimedia-based consumer electronics products, such as digital and cell-phone cameras, and intelligent video surveillance systems [20, 55, 150]. Of course, to understand such complex systems and the tools required to implement their algorithms, it is necessary to start with the basics.

Let us begin by examining the underlying concepts that form the foundation of such real-time systems. Starting with an analysis of the basic types of operations that are commonly encountered in image and video processing algorithms, it is argued that the real-time processing needs can be met through exploitation of various types of parallelism inherent in such algorithms. In what follows, the concept of "real-time" as it pertains to image and video processing systems is discussed and followed by an overview of the history of these systems and a glance at some of the emerging applications along with the common types of implementation trade-off decisions. This introductory chapter ends with a brief overview of the other chapters.

1.2   PARALLELISM IN IMAGE/VIDEO PROCESSING OPERATIONS

Real-time image and video processing systems involve processing vast amounts of image data in a timely manner for the purpose of extracting useful information, which could mean anything from obtaining an enhanced image to intelligent scene analysis. Digital images and video are essentially multidimensional signals and are thus quite data intensive, requiring a significant amount of computation and memory resources for their processing [15]. For example, take a typical $N \times M$ digital image frame with P bits of precision. Such an image contains $N \times M \times P$ bits of data. Normally, each pixel can be sufficiently represented as 1 byte or 8 bits, the exception being in medical or scientific applications where 12 or more bits of precision may be needed for higher levels of accuracy. The amount of data increases if color is also considered. Furthermore, the time dimension of digital video demands processing massive amounts of data per second. One of the keys to real-time algorithm development is the exploitation of the information available in each dimension. For digital images, only the spatial information can be exploited, but for digital videos, the temporal information between image frames in a sequence can be exploited in addition to the spatial information.

A common theme in real-time image/video processing systems is how to deal with their vast amounts of data and computations. For example, a typical digital video camera capturing VGA-resolution quality, color video (640×480) at 30 fps requires performing several stages of processing, known as the image pipeline, at a rate of 27 million pixels per second. Consider that in the near future as high-definition TV (HDTV) quality digital video cameras come into the market, approximately 83 million pixels per second must be processed for 1280×720 HDTV quality video at 30 fps. With the trend toward higher resolution and faster frame rates, the amounts of data that need to be processed in a short amount of time will continue to increase dramatically.

The key to cope with this issue is the concept of parallel processing, a concept well known to those working in the computer architecture area, who deal with computations on large data sets. In fact, much of what goes into implementing an efficient image/video processing system centers on how well the implementation, both hardware and software, exploits different forms of parallelism in an algorithm, which can be data level parallelism (DLP) or/and instruction level parallelism (ILP) [41, 65, 134]. DLP manifests itself in the application of the same operation on different sets of data, while ILP manifests itself in scheduling the simultaneous execution of multiple independent operations in a pipeline fashion.

To see how the concept of parallelism arises in typical image and video processing algorithms, let us have a closer look at the operations involved in the processing of image and video data. Traditionally, image/video processing operations have been classified into three main levels, namely low, intermediate, and high, where each successive level differs in its input/output data relationship [41, 43, 89, 134]. Low-level operators take an image as their input and produce an image as their output, while intermediate-level operators take an image as their input and generate image attributes as their output, and finally high-level operators

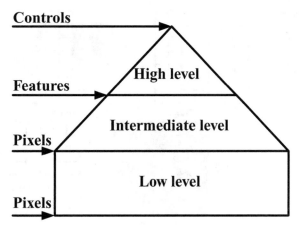

FIGURE 1.1: Image processing operations pyramid

take image attributes as their inputs and interpret the attributes, usually producing some kind of knowledge-based control at their output. As illustrated in Figure 1.1, this hierarchical classification can be depicted as a pyramid with the pixel data intensive operations at the bottom level and the more control-intensive, knowledge-based operations at the top level with feature extraction operations in-between the two at the intermediate level. Each level of the pyramid is briefly explained here, revealing the inherent DLP in many image/video processing operations.

1.2.1   Low-Level Operations

Low-level operations transform image data to image data. This means that such operators deal directly with image matrix data at the pixel level. Examples of such operations include color transformations, gamma correction, linear or nonlinear filtering, noise reduction, sharpness enhancement, frequency domain transformations, etc. The ultimate goal of such operations is to either enhance image data, possibly to emphasize certain key features, preparing them for viewing by humans, or extract features for processing at the intermediate-level.

These operations can be further classified into point, neighborhood (local), and global operations [56, 89, 134]. Point operations are the simplest of the low-level operations since a given input pixel is transformed into an output pixel, where the transformation does not depend on any of the pixels surrounding the input pixel. Such operations include arithmetic operations, logical operations, table lookups, threshold operations, etc. The inherent DLP in such operations is obvious, as depicted in Figure 1.2(a), where the point operation on the pixel shown in black needs to be performed across all the pixels in the input image.

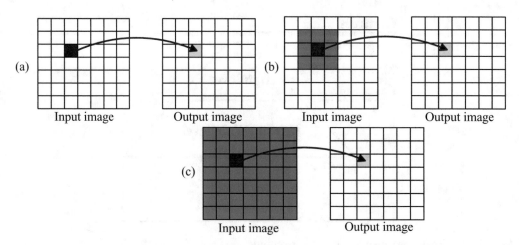

FIGURE 1.2: Parallelism in low-level (a) point, (b) neighborhood, and (c) global image/video processing operations

Local neighborhood operations are more complex than point operations in that the transformation from an input pixel to an output pixel depends on a neighborhood of the input pixel. Such operations include two-dimensional spatial convolution and filtering, smoothing, sharpening, image enhancement, etc. Since each output pixel is some function of the input pixel and its neighbors, these operations require a large amount of computations. The inherent parallelism in such operations is illustrated in Figure 1.2(b), where the local neighborhood operation on the pixel shown in black needs to be performed across all the pixels in the input image.

Finally, global operations build upon neighborhood operations in which a single output pixel depends on every pixel in the input image [see Figure 1.2(c)]. A prominent example of such an operation is the discrete Fourier transform which depends on the entire image. These operations are quite data intensive as well.

All low-level operations involve nested looping through all the pixels in an input image with the innermost loop applying a point, neighborhood, or global operator to obtain the pixels forming an output image. As such, these are fairly data-intensive operations, with highly structured and predictable processing, requiring a high bandwidth for accessing image data. In general, low-level operations are excellent candidates for exploiting DLP.

1.2.2   Intermediate-Level Operations

Intermediate-level operations transform image data to a slightly more abstract form of information by extracting certain attributes or features of interest from an image. This means that such operations also deal with the image at the pixel level, but a key difference is that the trans-formations involved cause a reduction in the amount of data from input to output. Intermediate operations primarily include segmenting an image into regions/objects of interest, extracting edges, lines, contours, or other image attributes of interest such as statistical features. The goal by carrying out these operations is to reduce the amount of data to form a set of features suitable for further high-level processing. Some intermediate-level operations are also data intensive with a regular processing structure, thus making them suitable candidates for exploiting DLP.

1.2.3   High-Level Operations

High-level operations interpret the abstract data from the intermediate-level, performing high-level knowledge-based scene analysis on a reduced amount of data. Such operations include classification/recognition of objects or a control decision based on some extracted features. These types of operations are usually characterized by control or branch-intensive operations. Thus, they are less data intensive and more inherently sequential rather than parallel. Due to their irregular structure and low-bandwidth requirements, such operations are suitable candidates for exploiting ILP [20], although their data-intensive portions usually include some form of matrix–vector operations that are suitable for exploiting DLP.

1.2.4   Matrix–Vector Operations

It is important to note that in addition to the operations discussed, another set of operations is also quite prominent in image and video processing, namely matrix–vector operations. Linear algebra is used extensively in image and video processing, and most algorithms require at least some form of matrix or vector operations, even in the high-level operations of the processing chain. Thus, matrix–vector operations are prime candidates for exploiting DLP due to the structure and regularity found in such operations.

1.3   DIVERSITY OF OPERATIONS IN IMAGE/VIDEO PROCESSING

From the above discussion, one can see that there is a wide range of diversity in image and video processing operations, starting from regular, high data rate operations at the front end and proceeding toward irregular, low data rate, control-intensive operations at the back end [1]. A typical image/video processing chain combines the three levels of operations into a complete system, as shown in Figure 1.3, where row (a) shows the image/video processing chain, and

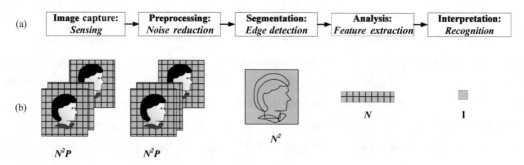

FIGURE 1.3: Diversity of operations in image/video processing: (a) typical processing chain, (b) decrease in amount of data across processing chain

row (b) shows the decrease in the amount of data from the start of the chain to the end for an $N \times N$ image with P bits of precision [126].

Depending on the types of operations involved in an image/video processing system, this leads to the understanding that a single processor might not be suitable for implementing a real-time image/video processing algorithm. A more appropriate solution would thus involve a highly data parallel front end coupled with a fast general-purpose back end [1].

1.4 DEFINITION OF "REAL-TIME"

Considering the need for real-time image/video processing and how this need can be met by exploiting the inherent parallelism in an algorithm, it becomes important to discuss what exactly is meant by the term "real-time," an elusive term that is often used to describe a wide variety of image/video processing systems and algorithms. From the literature, it can be derived that there are three main interpretations of the concept of "real-time," namely real-time in the perceptual sense, real-time in the software engineering sense, and real-time in the signal processing sense.

1.4.1 Real-time in Perceptual Sense

Real-time in the perceptual sense is used mainly to describe the interaction between a human and a computer device for a near instantaneous response of the device to an input by a human user. For instance, Bovik [15] defines the concept of "real-time" in the context of video processing, describing that *"the result of processing appears effectively 'instantaneously' (usually in a perceptual sense) once the input becomes available."* Also, Guy [60] defines the concept of *"real-time image processing"* as the *"digital processing of an image which occurs seemingly immediately; without a user-perceivable calculation delay."* An important item to observe here is that "real-time" involves the interaction between humans and computers in which the use of the words "appears" and "perceivable" appeals to the ability of a human to sense delays. Note that "real-time" connotes the idea of a maximum tolerable delay based on human perception of delay, which is essentially some sort of application-dependent bounded response time.

For instance, the updating of an automatic white balance (AWB) algorithm running on a digital camera need not operate every 33 ms at the maximum frame rate of 30 fps. Instead, updating at approximately 100 ms is sufficient for the processing to seem imperceptible to a human user when white balance gains require adjustment to reflect the surrounding lighting conditions. Thus, as long as the algorithm takes no longer than 100 ms to complete whatever image processing the algorithm entails, it can be considered to be "real-time." It should be noted that in this example, in certain instances, for example low-light conditions, it might be perfectly valid to relax the "real-time" constraint and allow for extra processing in order to achieve better image quality. The key question is whether an end user would accept the trade-off between slower update rates and higher image quality. From this discussion, one can see that the definition of "real-time" is loose because the maximum tolerable delay is entirely application dependent and in some cases the system would not be deemed a complete failure if the processing happened to miss the "real-time" deadline.

1.4.2   Real-time in Software Engineering Sense

Real-time in the software engineering sense is also based on the concept of a bounded response time as in the perceptual sense. Dougherty and Laplante [42] point out that a *"real-time system is one that must satisfy explicit bounded response time constraints to avoid failure,"* further explaining that *"a real-time system is one whose logical correctness is based both on the correctness of the outputs and their timeliness."* Indeed, while any result of processing that is not logically correct is useless, the important distinction for "real-time" status is the all-important time constraint placed on obtaining the logically correct results.

In software engineering, the concept of "real-time" is further classified based on the strictness attached to the maximum bounded response time into what is known as hard real-time, firm real-time, and soft real-time. Hard real-time refers to the case where if a real-time deadline is missed, it is deemed to be a complete failure. Firm real-time refers to the case in which a certain amount of missed real-time deadlines is acceptable and does not constitute failure. Finally, soft real-time refers to the case where missed real-time deadlines result in performance degradation rather than failure. In order to manage the priorities of different tasks of a system, real-time operating systems have been utilized to ensure that deadlines, whether hard, firm, or soft, are met. From a software engineer point of view, the issue of real-time is more about predictable performance rather than just fast processing [90].

1.4.3   Real-time in Signal Processing Sense

Real-time in the signal processing sense is based on the idea of completing processing in the time available between successive input samples. For example, in [81], "real-time" is defined as "*completing the processing within the allowable or available time between samples,*" and it is stated that a real-time algorithm is one whose total instruction count is "*less than the number of instructions that can be executed between two consecutive samples.*" While in [1], "real-time processing" is defined as the computation of "*a certain number of operations upon a required amount of input data within a specified interval of time, set by the period over which the data arrived.*" In addition to the time required for processing, the times required for transferring image data and for other memory-related operations pose additional bottlenecks in most practical systems, and thus they must be taken into consideration [124].

An important item of note here is that one way to gauge the "real-time" status of an algorithm is to determine some measure of the amount of time it takes for the algorithm to complete all requisite transferring and processing of image data, and then making sure that it is less than the allotted time for processing. For example, in multimedia display devices, screen updates need to occur at 30 fps for humans to perceive continuous motion, and thus any picture enhancement or other types of image/video processing must occur within the 33 ms time frame. It should be pointed out that, in image/video processing systems, it is not always the case that the processing must be completed within the time afforded by the inverse frame rate, as was seen in the above AWB update example.

1.4.4   Misinterpretation of Concept of Real-time

A common misunderstanding regarding the concept of "real-time" is that since hardware is getting faster and more powerful each year, "real-time" constraints can be met simply by using the latest, fastest, most powerful hardware, thus rendering "real-time," a nonissue. The problem with this argument is that it is often the case that such a solution is not a viable one, especially for consumer electronics embedded systems that have constraints on their total system cost and power consumption. For instance, it does not make sense to bundle the engineering workstation used to develop an image processing algorithm into a digital camera just for the purpose of running the algorithm in real-time.

1.4.5   Challenges in Real-time Image/Video Processing

Bearing in mind the above argument, developing a real-time image/video processing system can be quite a challenge. The solution often ends up as some combination of hardware and software approaches. From the hardware point of view, the challenge is to determine what kind of hardware platform is best suited for a given image/video processing task among the myriad of available hardware choices. From the algorithmic and/or software point of view, the challenge involves being able to guarantee that "real-time" deadlines are met, which could involve making choices between different algorithms based on computational complexity, using a real-time operating system, and extracting accurate timing measurements from the entire system by profiling the developed algorithm.

1.5   HISTORICAL PERSPECTIVE

The development of digital computers, electronic image sensors coupled with analog-to-digital converters, along with the theoretical developments in the field of multidimensional signal processing have all led to the creation of the field of real-time image and video processing. Here, an overview of the history of image processing is stated in order to gain some perspective on where this field stands today.

1.5.1   History of Image/Video Processing Hardware Platforms

The earliest known digital image processing, the processing of image data in digital form by a digital computer, occurred in 1957 with the first picture scanner attached to the National Bureau of Standards Electronic Automatic Computer (SEAC), built and designed by the scientists at the United States National Bureau of Standards, now known as the National Institute of Standards and Technology [86]. This scanner was used to convert an analog image into discrete pixels, which could be stored in the memory of the SEAC. The SEAC was used for early experiments in image enhancement utilizing edge enhancement filters. These developments, stimulated by the search for innovative uses of the ever-increasing computation power of computers, eventually led to the creation of the field of digital image processing as it is known today.

Around the same time frame, in the 1960s, developments at NASA's Jet Propulsion Laboratory led to the beginning of electronic imaging using monochrome charge-coupled device enabled electronic still cameras [56]. The need for obtaining clear images from space exploration was the driving force behind the uses of digital cameras and digital image processing by NASA scientists.

With such technology at hand, new applications for image processing were quickly developed, most notably including among others, industrial inspection and medical imaging. Of course, due to the inherent parallelism in the commonly used low-level and intermediate level operations, architectures for image processing were built to be massively parallel in order to cope with the vast amounts of data that needed to be processed. While the earliest computers used for digital processing of images consisted of large, parallel mainframes, the drive for miniaturization and advancements in very large scale integration (VLSI) technology led to the arrival of small, power-efficient, cost-effective, high-performance processor solutions, eventually bringing the processing power necessary for real-time image/video processing into a device that could fit in the palm of one's hand and go into a pocket.

It used to be that when an image/video system design required a real-time throughput, multiple boards with multiple processors working in parallel were used, especially in military and medical applications where in many cases cost was not a limiting factor. With the development of the programmable digital signal processor (DSP) technology in the 1980s though, this way of thinking was about to change. The following decade saw the introduction of the first commercially available DSPs, which were created to accelerate the computations necessary for signal processing algorithms. DSPs helped to usher in the age of portable embedded computing.

The mid-1980s also saw the introduction of programmable logic devices such as the field programmable gate array (FPGA), a technology that desired to unite the flexibility of software through programmable logic with the speed of dedicated hardware such as application-specific integrated circuits. In the 1990s, there was further growth in both DSP performance, through increased use of parallel processing techniques, and FPGA performance to meet the needs of multimedia devices and a push toward the concept of system-on-chip (SoC), which sought to bring all necessary processing power for an entire system onto a single chip. The trend for SoC design continues today [71].

In addition to these developments, a recent trend in the research community has been to harness the massive parallel computation power of the graphics processing units (GPUs) found in most modern PCs and laptops for performing compute-intensive image/video processing algorithms [110]. Currently, GPUs are used only in desktops or laptops, but pretty soon they are expected to be found in embedded devices as well. Another recent development that started in the late 1990s and early 2000s is the idea of a portable multimedia supercomputer that combines the high-performance parallel processing power needed by low-level and intermediate level image/video operations with the high energy efficiency demanded by portable embedded devices [54].

1.5.2   Growth in Applications of Real-time Image/Video Processing

Alongside the developments in hardware architectures for image/video processing, there have also been many notable developments in the application of real-time image/video processing. Lately, digital video surveillance systems have become a high-priority topic of research worldwide [6, 16, 36, 37, 40, 45, 69, 98, 149, 155]. Relevant technologies include automatic, robust face recognition [11, 28, 92, 112, 146], gesture recognition [111, 142], tracking of human or object movement [9, 40, 61, 68, 76, 92, 102, 151], distributed or networked video surveillance with multiple cameras [17, 37, 53, 75], etc. Such systems can be categorized as being hard real-time systems and require one to address some difficult problems when deployed in real-world environments with varying lighting conditions. Along similar lines, the development of smart camera systems [20] can be mentioned, which have many useful applications such as lane change detection warning systems in automobiles [133], monitoring driver alertness [72], or intelligent camera systems that can accurately adjust for focus [52, 79, 115, 116], exposure [13, 78, 108], and white balance [30, 78, 108] in response to a changing scene. Other interesting areas of research include developing fast, efficient algorithms to support the image/video coding standards set forth by the standards committees [22, 26, 31, 33, 48, 57, 70, 73, 82, 87, 106, 144]. In the never ending quest for a perfect picture, research in developing fast, high-quality algorithms for processing pictures/videos captured by consumer digital cameras or cell-phone cameras [80] is expected to continue well into the future. Of course, the developments in industrial inspection [25, 34, 67, 135, 147] and medical imaging systems [18, 23, 24, 44, 136, 143, 145] will continue to progress. The use of color image data [8, 85, 107, 109], or in some cases, multispectral image data [139] in real-time image/video processing systems is also becoming an important area of research.

It is worth mentioning that the main sources of inspiration for all the efforts in the applications of real-time image/video processing are biological vision systems, most notably the human visual system. As Davies [35] puts it, "if the eye can do it, so can the machine." This requires using our knowledge along with the available algorithmic, hardware, and software tools to properly transition algorithms from research to reality.

1.6   TRADE-OFF DECISIONS

Designing real-time image/video processing systems is a challenging task indeed. Given a fixed amount of hardware, certain design trade-offs will most certainly have to be made during the course of transitioning an algorithm from a research development environment to an actual real-time operation on some hardware platform. Practical issues of speed, accuracy, robustness, adaptability, flexibility, and total system cost are important aspects of a design and in practice, one usually has to trade one aspect for another [35]. In real-time image/video processing systems, speed is critical and thus trade-offs such as speed versus accuracy are commonly encountered. Since the design parameters depend on each other, the trade-off analysis can be viewed as a system optimization problem in a multidimensional space with various constraint curves and surfaces [35]. The problem with such an analysis is that, from a mathematical viewpoint, methods to determine optimal working points are generally unknown, although some progress is being made [62]. As a result, one is usually forced to proceed in an *ad hoc* manner.

1.7   CHAPTER BREAKDOWN

It could be argued that we are at a crossroad in the development of real-time image/video processing systems. Although high-performance hardware platforms are available, it is often difficult to easily transition an algorithm onto such platforms.

The advancements in integrated circuit technology have brought us to the point where it is now feasible to put into practical use the rich theoretical results obtained by the image processing community. The value of an algorithm hinges upon the ease with which it can be placed into practical use. While the goal of implementing image/video processing algorithms in real-time is a practical one, the implementation challenges involved have often discouraged researchers from pursuing the idea further, leaving it to someone else to discover the algorithm, explore its trade-offs, and implement a practical version in real-time. The purpose of the following chapters is to ease the burden of this task by providing a broad overview of the tools commonly used in practice for developing real-time image/video processing systems. The rest of the book is organized as follows:

- **Chapter 2:** *Algorithm Simplification Strategies*
  In this chapter, the algorithmic approaches for implementing real-time image/video processing algorithms are presented. It includes guidelines as how to speed up commonly used image/video processing operations. These guidelines are gathered from the recent literature spanning over the past five years.

- **Chapter 3:** *Hardware Platforms for Real-Time Image and Video Processing*
  In this chapter, the hardware tools available for implementing real-time image/video processing systems are presented, starting from a discussion on what kind of hardware is needed for a real-time system and proceeding through a discussion on the processor options available today such as DSPs, FPGAs, media-processor SoCs, general-purpose

processors, and GPUs, with references to the recent literature discussing such hardware platforms.

- **Chapter 4:** *Software Methods for Real-Time Image and Video Processing*
  This chapter covers the software methods to be deployed when implementing real-time image/video processing algorithms. Topics include a discussion on software architecture designs followed by a discussion on memory and code optimization techniques.

- **Chapter 5:** *The Road Map*
  The book culminates with a suggested methodology or road map for the entire process of transitioning an algorithm from a research development environment to a real-time implementation on a target hardware platform using the tools and resources mentioned throughout the previous three chapters.

CHAPTER 2

Algorithm Simplification Strategies

2.1 INTRODUCTION

An algorithm is simply a set of prescribed rules or procedures that are used to solve a given problem [103, 130]. Although there may exist different possible algorithms for solving an image/video processing problem, when transitioning to a real-time implementation, having efficient algorithms takes higher precedence. Efficiency implies low computational complexity as well as low memory and power requirements. Due to the vast amounts of data associated with digital images and video, developing algorithms that can deal with such amounts of data in a computational, memory, and power-efficient manner is a challenging task, especially when they are meant for real-time deployment on resource constrained embedded platforms.

Since algorithms are usually prototyped in development environments not suffering from resource constraints, they often have to be "optimized" for achieving real-time performance on a given hardware platform. While special hardware and software optimization techniques can be used to realize a real-time version of an algorithm, in general, greater gains in performance are obtained through simplifications at the algorithmic level [1, 124]. Such modifications or simplifications performed at the algorithmic level help to streamline the algorithm down to its core functionality, which not only leads to a lower computational complexity but also to lower memory and power requirements.

Thus, the very first step in transitioning an algorithm from a research environment to a real-time environment involves applying simplification strategies to the algorithm. It is more effective to perform these simplifications while still working in the research development environment, which possesses a higher design flexibility over the implementation environment. As the first step toward transitioning an algorithm to a real-time implementation, this chapter presents the strategies to achieve algorithmic simplifications along with relevant examples from the literature to exhibit successful applications of the strategies.

2.2   CORE SIMPLIFICATION CONCEPTS

Examining the literature on real-time image and video processing reveals three major concepts addressing algorithmic simplifications. These strategies include the following:

- reduction in number of operations;
- reduction in amount of data to be processed; and
- utilization of simple or simplified algorithms.

Reductions in operations and in amount of data to be processed are the most common simplification strategies for transitioning an algorithm to a real-time implementation. While there are many different manifestations of the reduction strategies, the discussion here provides a general classification of the strategies that are scattered throughout the literature. Each of these concepts will be expanded upon in the following subsections.

2.2.1   Reduction in Number of Operations

Due to the enormous amount of data involved in image/video processing, every operation counts, especially the time-consuming operations at the lower levels of the processing hierarchy. Thus, reduction in the number of operations plays a major role in achieving real-time performance.

2.2.1.1 Pure Operation Reduction

The strategy of pure operation reduction involves applying a transformation to reduce the number of operations, which does not change the numerical outcome. If the numerical outcome is changed, then the approach is referred to as either an approximation or a suboptimal/alternative solution as discussed in the next subsection. Any operation that has inherent symmetries or redundancies in its constituent computations is a candidate for applying this strategy. Application of this strategy manifests itself in uncovering hidden symmetries or redundancies within the computations, which can often be discovered by expanding the computations by hand and carefully noting any mathematical identities or properties [1].

An example of the reduction in number of operations through inherent computation symmetries or redundancies can be found in sliding window operations. Oftentimes, the extraction of local statistics via sliding windows involves a heavy amount of redundant computations that can be reduced by formulating a recursive approach. A common technique is to keep a running accumulator, where new results are obtained by subtracting the results of the leftmost column of the sliding window from the accumulator, sliding the window to the right to the next position, and adding the result of the new set of pixels in the rightmost column of the window to the accumulator. Another example of reduction in the number of operations arises in linear filtering utilizing a specific filter kernel. Symmetries in kernel coefficients can be used to reduce the amount of computations without changing the numerical outcome. Reduction of computations can also be achieved by rearranging or regrouping the computations such that the occurrence of the more time-consuming operations such as multiplication and division is reduced.

Throughout the literature, time-consuming operations including multiplication and division as well as other so-called "expensive" operations are considered the main targets for reduction. These operations are regarded as "expensive" due to the fact that they usually require more time to execute than other operations such as addition or bit-shifting. Thus, they present obstacles to achieving real-time performance. Researchers often seek to restructure an algorithm in such a way to reduce the amount of multiplications or divisions, sometimes even replacing these operations with simple bit-shifting to carry out multiplications or divisions by powers of 2. The use of computationally simple operations is one of the key steps toward achieving real-time performance. Other times, a calculation can be cleverly factored, causing a reduction in the number of operations [12].

2.2.1.2 Operation Reduction Through Approximations

The strategy of approximations is similar to the strategy of reduction in computations in that approximations involve applying transformations to reduce the number of operations, but it differs from pure computational reduction due to the presence of approximation errors. The main objective of this strategy is to minimize errors as much as possible, thus obtaining suboptimal results within an acceptable interval of time. Consequently, in this strategy, one seeks a trade-off between the accuracy of the outcome and the speed at which it is obtained. In fact, a real-time performance is often obtained by trading off accuracy of processing with speed of processing, where an "optimal" working point is found through experimentation. Examples of this strategy include the use of lookup tables for computationally complex calculations, the use of a suboptimal search technique over a time-consuming, exhaustive search technique, the use of sums of absolute values over time-consuming squaring operations, and many others.

2.2.1.3 Alternative Method

Choosing an alternative method is a strategy that involves searching for a new algorithm in order to attain faster processing while at the same time maintaining a required level of accuracy. This strategy is similar to the strategy of approximations, except that it is used primarily when the algorithm is too computationally complex for any approximations or reductions to yield an acceptable real-time performance. Basically, this strategy involves abandoning the computationally complex algorithm in favor of one that is less computationally complex. A simpler algorithm is often developed by careful consideration of data to be processed and properties of the objects being sought out.

2.2.2   Reduction in Amount of Data

In addition to the reduction of operations, the reduction of data to be processed also plays a major role toward having real-time performance. This is simply because such systems require the processing of large amounts of data in relatively short intervals of time, and thus any reduction in the amount of data can lead to a reduction in the processing time. This strategy involves applying a transformation to the underlying image data for the purpose of deriving a compact representation, which in turn speeds up subsequent stages of processing. Reduction of data takes many forms in real-time image/video processing systems including spatial or temporal down-sampling, spatial block partitioning, region of interest or selective processing, formulating the algorithm in a multiresolution or coarse-to-fine processing framework, appropriate feature extraction, etc. In all these cases, a certain subset of pixels from an image frame is processed.

2.2.2.1   Spatial Down-Sampling, Selective Processing, and Coarse-to-fine Frameworks

Spatial down-sampling involves skipping the processing of a certain amount of pixels within a frame, while temporal down-sampling involves skipping the processing of an entire frame within a video sequence. It should be noted that such down-sampling may not be applicable in certain applications. Spatial block partitioning involves dividing up an image frame into nonoverlapping or overlapping blocks, and separately processing each block. Region of interest or selective processing involves applying complex computations only to a desired subset of the image data, providing a speedup in processing by narrowing down the region of interest. A multiresolution or coarse-to-fine framework involves formulating an algorithm where a rough estimate is first obtained at a coarse resolution level and then subsequently refined at increasing levels of resolution to allow a rapid convergence to the solution.

2.2.2.2   Compact Representation Through Appropriate Features

Selection of appropriate features to represent objects of interest helps one to cut down the amount of extraneous information, providing a succinct representation of information which in turn simplifies higher level operations such as Kalman filter tracking. Effective features depend on the application and the objects involved. For instance in face detection applications, color has been found to be an effective feature for real-time deployment. Of course, there is a computational burden associated with the extraction of features from image/video sequences. Hence, another important aspect involves not only finding effective features but also fast mechanisms to compute them.

Automatic algorithms that require less human intervention and fewer or no user-defined thresholds or parameters are also preferred because such algorithms can adapt to different situations automatically, leading to a truly stand-alone real-time system, which is the ultimate goal of most practical designs [6]. Clearly, the adaptation schemes should be kept computationally simple and not burden the rest of the algorithm. In addition to this, since real-time image/video processing systems are usually employed in real-world environments, in many cases, data pre-processing is performed to remove unwanted disturbances such as noise to ensure generating a reliable outcome. Similar to any other component, the preprocessing should also be as computationally efficient as possible.

2.2.3 Simplified Algorithms

In general, the fundamental idea behind real-time image/video processing systems is the utilization of simple or simplified algorithms. A rule of thumb when transitioning to a real-time implementation is to keep things as simple as possible, that is to say, look for simple solutions using simple operations and computationally simple algorithms as opposed to complex, computationally intensive algorithms, which may be optimal from a mathematical viewpoint, but are not practical from a real-time point of view. With embedded devices now being outfitted with vision capabilities, such as camera-equipped cell phones and digital still/video cameras, the deployment of those algorithms which are not only computationally efficient but also memory efficient is expected to grow. Often, algorithms are carefully analyzed in terms of their number of operations and computational complexity, but equally important are their storage requirements. In essence, simple algorithms provide a means of meeting real-time performance goals by lowering computational burdens, memory requirements, and indirectly, power requirements as well.

2.3 EXAMPLES OF SIMPLIFICATIONS

To help illustrate the three stated algorithm simplification strategies toward achieving a real-time implementation, the following subsections present representative examples from the literature, which exhibit successful applications of the strategies.

2.3.1   Reduction in Number of Operations

2.3.1.1 Pure Reduction

The strategy of pure operation reduction has been used by many researchers to simplify algorithms for real-time implementation. This strategy has been primarily used for algorithms with repetitive, well-structured computations in low-level operations, such as filtering, transforms, matrix–vector operations, and local statistics extraction. This subsection includes several examples illustrating the strategy of pure operation reduction.

Exploiting any available symmetry in the computations involved can often lead to pure operation reduction. For instance, in [87], the symmetry of the coefficients of linear-phase filters in the biorthogonal wavelet transform allowed streamlining the computation of the forward and inverse transforms into a single architecture while reducing the number of expensive multiplication operations. This produced a more efficient transform computation. Other common techniques used for reducing the number of multiplication operations in linear filtering include making use of the separability of the kernel involved, or eliminating multiplications by ones or zeros [99].

Computations can often be cleverly rearranged or factored to reduce the number of operations. One example of this can be found in [106], where the symmetry in the elements of the discrete cosine transform (DCT) matrix allowed rearranging the computations, reducing the number of expensive multiplication as well as addition operations. As a result, a more efficient transform was achieved through this operation reduction.

Another possible way to achieve pure operation reduction in matrix computations encountered in image/video processing algorithms is to seek out encoding schemes that can transform a matrix into a sparse matrix. For example, in [48], an exact two-dimensional (2D) polynomial expansion in terms of integer coefficients provided a sparse representation of the elements of the DCT matrix. This allowed a more efficient computation by replacing expensive multiplication operations with simple bit-shift and addition operations and reducing the number of multiplications as well as additions. Another popular technique to reduce the amount of operations in matrix computations involves exploiting matrix properties. For example, in [74], a rearrangement of the equations for a 2.5D affine motion parameter estimation allowed an efficient solution via an orthogonal matrix factorization using a Householder transformation, thus reducing the computation operations over that of a 2D affine estimation.

Many times, an efficient computational structure derived from digital signal processing theory can be utilized to achieve a reduction in the number of operations. For example, in [94], a one-dimensional (1D) infinite impulse response filter provided a reduction in the number of expensive multiplication operations per pixel over that of a 1D finite impulse response filter in addition to saving memory space via using a lower order filter. These changes led to an efficient scan-line-based image enhancement at video rates. Another example of this approach can be found in [88], where the relationship between the computation structure of discrete, geometric moments and that of all-pole digital filters was exploited, allowing the computation of any order geometric moment using a series of accumulators. This resulted in a significant reduction in the number of multiplications and thus allowed real-time computation of geometric moments.

In addition to multiplication operations, reduction in the number of comparison operations may also be useful. For instance, in [57], it was reported that the comparison operations had a greater computational cost than that of the addition and subtraction operations in the sum-of-absolute difference (SAD) computation for determining motion vectors as part of a block-matching method. In order to reduce the number of comparisons to determine the minimum of the SAD error surface, the order in which the SADs were calculated was modified from the standard scan-line order to a circular, spiraling inward order. This allowed comparing the SAD values only twice per spiral, providing up to a 14% decrease in comparison operations per frame with virtually no noticeable loss in image quality. Thus, changing the order by which a computation is carried out may lead to a reduction in the number of operations.

Exploiting recursion in a computation can be quite useful toward reducing the number of operations in certain data-intensive, low-level image processing tasks. This technique capitalizes on the redundancies present in adjacent computations. For example, in [84], a recursion in the computation of local variances was used to reduce the redundant computations by adding the values of a right strip and subtracting the values of a left strip. This provided a significant speedup from 112 to 3.28 ms, thus achieving the required real-time performance.

Similarly, in [7], a recursion in the correlation computation as part of a similarity measure for stereo image matching was used to reduce the amount of redundant computations for adjacent pixels in overlapping windows. A noteworthy discussion detailing the generalization of this approach to higher dimensions can be found in [137], where the 2D sliding window recursion equation was generalized to N-dimensional images. This recursion reduced the redundant operations in computing local statistics and provided a speedup that was independent of the size of the window. In [122], the strategy of recursion was applied to the computation of histogram statistics, where the computation of local histogram statistics was performed recursively and then generalized to N-dimensional images. This provided a remarkable speedup, 63 ms as compared to 2 min, leading to real-time color histogram matching.

2.3.1.2 Reduction via Approximations

When it comes to real-time implementation, sometimes sacrifices in accuracy have to be made in order to achieve the required performance. In fact, most algorithms that are transitioned to a real-time environment are simplified using various approximations.

Approximations are often used for reducing the computations in transform operations. For instance, in [33], approximating the DCT computations by using up to three biorthogonal matrices led to replacing expensive multiplication operations with bit-shifting and addition operations. The approximation could be varied between two or three matrices, producing trade-offs between speed and accuracy. A high-performance version was able to process one 8×8 image block using less than one processor clock cycle per pixel.

Approximating computations by utilizing simple operations is often used to reduce the amount of processing time. In [131], simplifying the computations for computing the gradient image via a nonlinear filter led to the speedup needed to achieve an efficient real-time implementation of the noise-robust image enhancement procedure. Two key algorithmic simplifications that were made included approximating the quadratic filter using a normalized squared-gradient computation and replacing the normalization of the filter output by a power of 2 division. The loss in accuracy due to the division operation was kept low by considering the maximum value of the filter output for different images in the test set. Another example that utilized this approach can be found in [63], where approximating 2D optical flow with two, simpler 1D flow computations provided a significant reduction in the processing time which allowed generating the motion model parameters in real-time. In order to account for the loss in accuracy due to the approximation, a fast iterative least-squares formulation was used to subsequently refine the approximation within ten iterations, providing a real-time throughput around 10–20 fps. Also, in [66], a subpixel edge detector was simplified by approximating a critical but expensive division operation via a simple iterative minimization procedure using integer arithmetic.

In some cases, the processing resources are so scarce that a desired computation cannot be fully realized without some simplifying approximation. For example, in [152], the computations required for applying a large 13×13 filter kernel had to be approximated by using two sequential passes of a smaller 7×7 filter kernel due to a lack of processing resources supporting larger kernels. The approximation did not have any detrimental effect on the outcome of the object tracking system under consideration.

Discarding computations to reduce the number of expensive operations is often used as a simple means of achieving algorithmic simplification through approximation. For instance, in [120], the computations involved in three-dimensional (3D) nonlinear filtering operations for the suppression of impulsive noise were simplified by using only a certain subset of the volume elements in the 3D filter kernel. This led to an acceptable processing speed while maintaining sufficient noise suppression characteristics. In [73], another application of discarding computations consisting of a partial decoding of an encoded image or video sequence was discussed. In order to provide a real-time performance for viewing MPEG-2-encoded DVD movies on low-resolution mobile handsets, the expensive full-decode operation was approximated by keeping only the DC and a certain number of AC coefficients, and discarding the rest. The number of discarded coefficients could be varied to achieve a desired speed versus quality trade-off. As an example, one of the presented simplifications used three AC coefficients and required only 7 multiplications and 10 additions versus 4096 multiplications and 4032 additions for the full decoding of one block.

Oftentimes, a function evaluation is approximated as a lookup table, essentially trading off the time spent in computations with the time spent in accessing values from the memory space. This can help one to achieve a real-time performance. One good example of this lookup table approach involves the computation of the entropy function. In [23], in order to reduce the error in the approximation, a multiresolution lookup table was considered using a higher resolution for the portion of the function that incurred a larger error.

Formulating an algorithm in an evolutionary manner may also provide different degrees of approximation. For example, in [133], in order to have a real-time tracking algorithm for an automobile obstacle avoidance warning system, the matching stage was approximated by formulating it in an evolutionary manner where the matching improved by increasing the processing time. Formulating an algorithm in this manner allows the simplification in computations to be performed according to the time made available for processing and thus ensures an acceptable outcome while meeting hard real-time constraints.

2.3.1.3 Reduction via Alternative Methods

Sometimes an algorithm may be too computationally complex to warrant exploration of computational reduction strategies. In such cases, it is often desired to develop a computationally simple algorithm that can reduce the complexity while at the same time maintain the high level of accuracy of the original algorithm. In [29], one good example was presented that involved the development of an efficient line detection algorithm. In this work, it was decided to abandon the use of computationally expensive Hough transform techniques in favor of a simpler algorithm by exploiting the underlying geometry of the objects under consideration. This alternative, simple algorithm was based on the idea that the area formed by three colinear points is zero. This geometric property allowed utilizing a computationally simple means of detecting lines in images, which led to a real-time deployment. Another example of using an alternative, simple algorithm that achieved a real-time performance appears in [111]. In this algorithm, the operation of fitting an ellipse to extract contour points for a human gesture tracking system was found to be the bottleneck. It was decided to replace the original algorithm, which required an expensive Levenberg–Marquardt-based fitting procedure, with an alternative moments-based algorithm. This replacement produced a significant speedup in the ellipse fitting procedure. Thus, sometimes it is beneficial to consider alternative formulations of an algorithm to address its real-time performance bottlenecks.

2.3.2   Reduction of Data

Reduction in the amount of data to be processed plays a prominent role for bringing image/video processing algorithms into real-time, and there are many examples given in the literature showing how to use this strategy to simplify an algorithm toward reaching a real-time performance. Reducing the amount of data upfront allows one to speed up subsequent stages of processing. Also, it indirectly allows one to use more complex processing during subsequent stages of processing.

2.3.2.1 Subsampling

One of the simplest and most effective approaches involves applying spatial or temporal subsampling of the incoming image frame or video sequence. The objective here is to reduce the amount of data to be processed and thus to obtain a speedup for subsequent stages of processing. There are many examples involving spatial subsampling. For example, in [25], subimages were reduced in size to 20×20 pixels before being subjected to classification processing in order to meet the hard real-time performance demands of an industrial inspection system. In another application involving obstacle detection, the input image was subsampled to reduce the size by a factor of 2, after having been spatially low-pass filtered for reducing the effects of aliasing. Spatial subsampling has also been found to be useful for speeding up face recognition systems on embedded platforms. In [146], to meet the real-time performance requirements, input images of size 320×240 were downscaled to 80×60 to help speed up the discussed face recognition system. Similarly, in [11], input images of size 640×480 were downsized to 128×128 to help speed up the subsequent processing.

Another benefit of spatial subsampling involves reducing the number of points to search for object detection in industrial inspection systems. In such applications, it has been shown that subsampling can provide a speedup up to 25 times [34]. Naturally, temporal subsampling can be applied to the processing of video sequences. For example, in [142], it was suggested that a complex range or scene depth computation could be performed for every nth incoming frame instead of every frame in order to meet the real-time requirement of a gesture tracking algorithm. Of course, it should be pointed out that such methods do impart a certain loss in accuracy due to the discarding of some of the original data. Hence, it is best to explore several different subsampling intervals to determine the right balance between speed and accuracy. Also of note here is that in some cases, spatial subsampling might not be appropriate. In such cases, one may formulate the problem as a scalable algorithm [62]. This approach involves creating a quality control module to manage the processing for achieving a certain level of output quality according to the availability of system resources.

2.3.2.2 Partitioning

Another simple method of reducing the amount of data to be processed involves partitioning an image frame into smaller subimages that can each be processed at a faster speed than the entire image frame. This is similar to the divide-and-conquer strategy where the problem is divided into several smaller problems that are easier to solve [103, 130]. The most popular partitioning schemes include row-wise, column-wise, and block-wise. A good example appearing in [144] covers the computation of the discrete biorthogonal wavelet transform (DBWT) for high-definition TV compression. Due to the large data set involved, real-time processing was not feasible without partitioning a frame into nonoverlapping, square subimages and calculating the DBWT separately on each subimage. In another application discussed in [47], the image frame was partitioned into vertical strips in order to process the image in parallel on multiple processors. This generated a real-time implementation of the *à trous* wavelet transform. In the face detection algorithm covered in [112] for an embedded device, the input image was partitioned into nonoverlapping subimages, the size of which was chosen to balance the gain in speed versus the accuracy of detection. In another example mentioned in [8], it was found that partitioning the input image into subimages enabled the real-time implementation of the most time-consuming portion of a color-quantization algorithm. Since edge artifacts could be present between adjacent subimages, an overlapping partitioning scheme can be employed, thus producing a trade-off between speed and artifact suppression [35]. In [31], an overlapped block-partitioning scheme was used to correct for processing across partition borders in a wavelet transform computation.

2.3.2.3 Selective Processing

Selective processing is another popular data reduction method that involves narrowing down the region of interest before applying any subsequent processing. As a result, only a certain subset of the entire image is processed, hence the name selective processing. For example, in [125], in locating an object of interest to subpixel accuracy, instead of applying subpixel calculations on the entire image first and then locating the object of interest, the location was first narrowed down to a specific area, after which appropriate computations were performed to refine this location to subpixel precision. In this case, narrowing down the area of interest helped to reduce the amount of data to be processed by the subpixel refinement stage, thus generating a real-time subpixel accurate object detection. Similarly, in another application involving the use of range data for gesture tracking [142], the range data was calculated only for selected regions of interest and not for the entire image. Therefore, if computationally complex processing cannot be avoided to save processing time, it is best to apply such processing only to the areas of interest and not to the entire image.

2.3.2.4 Determining and Using Appropriate Features

Determining and using appropriate image features instead of images themselves provide an effective way to reduce the dimensionality of image data to be processed, thus speeding up subsequent processing stages. Here, 11 examples exhibiting this approach are mentioned, each of which played a crucial role in achieving a real-time performance in their respective applications.

- In [40], *coarse motion features* were preferred over dense features since it was found that dense features provided unnecessary tracking accuracy for the video surveillance application under consideration while demanding too much processing time.

- In [102], *marker features*, which were physically placed on the objects to be tracked, were found to be appropriate features to obtain a real-time performance in the discussed constrained object tracking application. Such features allowed an object to be quickly identified by its unique marker features.

- In [93], *contour features* were seen to provide a more compact representation over mesh features, achieving a real-time performance in calculating deformation parameters for a 3D face modeling application.

- In [97], *Gabor wavelet features* were shown to provide a better baseline feature set over those from an eigen-decomposition method for a face detection application. The use of such features allowed the elimination of time-consuming pre- and postprocessing stages, cutting down on the amount of processing and thus leading to real-time performance.

- In [151], *the head contour* was found to be an appropriate feature to be used for tracking human heads that turned around, considering a lack of texture features for utilizing a texture-based head tracking. The use of such features allowed the construction of a fast algorithm for tracking the displacement of the contours between subsequent frames. This in turn led to real-time throughput.

- In [9], *motion and head shape features* were found to be appropriate for a head detection and tracking algorithm. The use of the circle shape to represent the shape of the head allowed utilization of a fast Hough transform algorithm for real-time performance.

- In [155], *features reflecting human visual perception* or features that humans can use to distinguish between a moving object and its shadows such as color and texture-based features were found to be the most appropriate features for the object tracking application under consideration.

- In [68], *contour features* were used to enable a real-time object tracking via capturing the essential shape information. It was shown that contour features could be efficiently represented using a radial-based representation, which required a small amount of computations to update.

- In [105], *distance-independent features* were considered to be the right kind of features for a real-time recognition application with objects placed at varying distances. These features allowed objects at varying distances to be represented by a fixed amount of pixels, which in turn helped to speed up the subsequent matching algorithm and to achieve real-time recognition.

- In [121], *DCT coefficients and motion vectors* were utilized to achieve a real-time video object segmentation using MPEG-encoded video sequences. The use of these features allowed the construction of a feature vector that succinctly captured the frequency-temporal characteristics of the image blocks, providing a 320:1 reduction in the amount of data for a typical 352×288 color video sequence. This in turn helped to construct an efficient algorithm for segmenting video objects without resorting to the computationally expensive full decoding.

- *Color features* have been found to be useful in many human detection and tracking applications over those obtained from gray-level images. For example, in [75], color-based features were preferred over gray-level-based features due to their robust tracking performance in the face of poor conditions such as sudden changes in lighting. Also, in [92] and [28], skin color features were found to be the most appropriate for face detection applications. To balance out the increase in the amount of data to be processed due to the use of color information, it has been recommended to subsample color images to allow for faster processing [75].

2.3.2.5 Dimensionality Reduction

After having determined appropriate features to use, it is often helpful to further reduce the amount of data by applying dimensionality reduction techniques, such as principal component analysis (PCA), linear discriminant analysis (LDA), or a Kohonen self-organizing map (SOM). There are many examples in the literature that employ such techniques. For instance, in [27], the Kohonen SOM neural network was used to reduce the dimension of a feature vector to help speed up the subsequent processing, while in [76], PCA was employed for the same purpose. In [139], a combination of PCA and LDA was used to reduce the amount of data, providing a better separation of classes and thus easing the task of the subsequent classifier. Also, in [109], the dimensionality of chrominance feature vectors extracted from a 2D chrominance histogram was reduced by modeling the histogram as a multivariable Gaussian distribution and then applying PCA to remove the correlation in the two components. This allowed the use of 1D marginal statistics of each component instead of 2D statistics.

Dimensionality reduction has also been used when dealing with color images. When dealing with color image data, researchers have often made use of normalized 2D color spaces as opposed to 3D color spaces to achieve real-time implementations. For instance, in the face recognition application discussed in [28], the 2D normalized r–g color space was used, saving computations by reducing the 3D color space by one dimension. Also, reducing a three-channel color image to a one-channel image can sometimes serve as a means of achieving a real-time performance through dimensionality reduction. For instance, in [141], a moment-preserving threshold scheme was used to reduce a three-channel color image to a one-channel image by taking into account the correlation amongst the three color channels.

2.3.2.6 Multiresolution or Coarse-to-fine Processing

The technique of casting a complex problem into an appropriate multiresolution or coarse-to-fine processing framework can often provide a reduction in data and computation through quick determination of a rough or low-resolution solution and its subsequent refinement at higher resolutions.

Optimization involving the global minimization of a parameter over some error surface arises often in image/video processing algorithms. To overcome difficulties in reaching the global minimum, the optimization can be performed using a multiresolution framework. The advantage of this framework is that at lower resolutions, the error surface is smoother and contains fewer local minima, allowing the use of fast methods to quickly find a rough solution. This solution can then be propagated to a higher resolution with a reduced search range around the location of the rough solution. Such a scheme reduces the amount of data used to reach a solution. While it can provide a fast means of reaching a reasonable solution, it does not guarantee convergence to the true global solution. Thus, this framework can be considered to provide a fast, but suboptimal solution—which can be regarded as another example of the speed versus accuracy trade-off often encountered in real-time image/video processing systems. One successful example of the multiresolution framework applied to the optimization of a functional can be found in [129], where such a framework was utilized to generate panoramic ultrasound images in real-time.

Similar to a multiresolution framework, coarse-to-fine processing involves formulating a problem by quickly determining a rough solution during a first processing pass and then subsequently refining it during a second processing pass. In [72], such a framework was developed for an object tracking application, where a Kalman filter was used to roughly determine and subsequently refine the position of an object of interest. The predicted location was used as a rough estimate, and the prediction uncertainty was used to limit the search area for refinement processing. This reduced the amount of data to be searched in every frame and thus led to real-time performance.

2.3.3 Simple Algorithms

In order to meet real-time requirements, many researchers divide problems into stages composed of computationally simple operations. Simple operations include the use of binary, morphological, frame differencing, and various other computationally efficient operations. Due to the fact that such algorithms are often employed in real-world situations, they often have to be made robust to noise sources or disturbances in the scene such as changes in lighting conditions, or low-contrast scenes. The algorithm used to achieve a robust performance must also be computationally simple, and preferably fully automatic with little human intervention in the setting of thresholds or parameters. In the formulation of such an algorithm, a trade-off analysis between speed and accuracy needs to be performed in order to find a balance between achieving real-time performance and robustness to disturbances in the scene. Of course, the use of appropriate features that are robust to disturbances can help. It should be noted that the construction of simple or simplified algorithms involves both the use of computational and data reduction simplification strategies mentioned in the previous sections. Four examples are presented next to illustrate the practical usefulness of such simple algorithms.

The tracking of objects within sequences of images has long been an important problem in image/video processing. For example, in [61], a simple algorithm for tracking an object based on template matching was presented. To provide a more robust operation to changes in object shape, a rather simple extended snake algorithm was employed. First, the object was extracted by using simple frame differencing and applying simple morphological closing on the result to merge homogenous regions. Then, a data reduction procedure was done by using an appropriate feature to succinctly represent the object and provide control points for the snake algorithm. The use of simple algorithms during each stage allowed achieving robust real-time tracking.

The increase in the practical applications of face detection and recognition has increased the interest in their real-time implementations. Such implementations are possible via the use of simple algorithms. For instance, in [92], a simple algorithm for face detection based on the use of skin color features was discussed. In order to make the detection algorithm robust to regions in the background with skin-like colors, a simplified method was developed. First, motion detection via a simple frame differencing operation was used to distinguish the human face from the background. Then, to reduce the noise in the difference image, a simple morphological opening operation was used. Finally, a simple labeling operation was used to determine the region where the face was located. Again, the use of simple algorithms during the various stages of the image processing chain allowed achieving robust real-time face detection.

Industrial inspection systems are known for their strict hard real-time deadlines, forcing developers to devise algorithms with simple operations. For example, in [147], a simple algorithm for the extraction of paper watermarks as part of a defect inspection system was introduced. Due to the real-time constraints involved, a simple morphological white top-hat operation was used to extract the watermarks. This simple operation was found to be not only fast, but also robust to disturbances such as uneven illumination and low-contrast watermarks. In this case, the morphological operations provided a means of obtaining real-time performance. Similarly, in [135], simple defect location and feature extraction algorithms were employed to meet the hard real-time constraints in an industrial inspection system. In order to speed up the location detection process, the image was first binarized to find a rough estimate of the location of the defect. This rough location was then propagated to the full-resolution image, and a region around this location was selected for feature extraction. A simple gray-level difference method was used to extract texture features from the narrowed down defect region. The use of relatively simple operations enabled the real-time detection of defects and extraction of features for the subsequent classification.

2.4   SUMMARY

In this chapter, the three major strategies for achieving algorithmic simplifications were covered including reduction of computations, reduction of data, and simple algorithm design. In addition, a wide range of examples from the literature was mentioned in order to further illustrate the use of such strategies in actual real-time image/video processing systems. When it comes to developing an efficient algorithm for real-time deployment, one should examine each stage of the algorithm according to the required real-time performance constraints, carefully seeing if any of the principles or techniques presented in this chapter can be applied to gain an improved performance. Ultimately, it is up to the developer to determine the best balance between speed and accuracy for the application under consideration. In summary, this chapter has provided insights into available algorithmic simplification choices that should be considered at the beginning of the real-time implementation process.

CHAPTER 3

Hardware Platforms for Real-Time Image and Video Processing

3.1 INTRODUCTION

A great deal of the present growth in the field of image/video processing is primarily due to the ever-increasing performance available on standard desktop PCs, which has allowed rapid development and prototyping of image/video processing algorithms. The desktop PC development environment has provided a flexible platform in terms of computation resources including memory and processing power. In many cases, this platform performs quite satisfactorily for algorithm development. The situation changes once an algorithm is desired to run in real-time. This involves first applying algorithmic simplifications as discussed in Chapter 2, and then writing the algorithm in a standard compiled language such as C, after which it is ported over to some target hardware platform. After the algorithmic simplification process, there are different possible hardware implementation platforms that one can consider for the real-time implementation. The selection of an appropriate hardware platform depends on the responses provided to the following questions:

- What are the important features of an image/video processing hardware platform?

- What are the advantages and disadvantages associated with different hardware platforms?

- What hardware platforms have previously been used for the real-time application under consideration?

- What kind of hardware platform is best suited for the real-time application under consideration?

These questions will be examined in the sections that follow in this chapter.

3.2   ESSENTIAL HARDWARE ARCHITECTURE FEATURES

As discussed in Chapter 1, practical image/video processing systems include a diverse set of operations from structured, high-bandwidth, data-intensive, low-level and intermediate-level operations such as filtering and feature extraction, to irregular, low-bandwidth, control-intensive, high-level operations such as classification. Since the most resource demanding operations in terms of required computations and memory bandwidth involve low-level and intermediate level operations, considerable research has been devoted to developing hardware architectural features for eliminating bottlenecks within the image/video processing chain, freeing up more time for performing high-level interpretation operations. While the major focus has been on speeding up low-level and intermediate level operations, there have also been architectural developments to speed up high-level operations.

From the literature, one can see there are three major architectural features that are essential to any image/video processing system, namely single instruction multiple data (SIMD), very long instruction word (VLIW), and an efficient memory subsystem. The concept of SIMD processing is a key architectural feature found in one way or another in most modern real-time image/video processing systems [20, 35, 65]. It embodies broadcasting a single instruction to multiple processors, which simultaneously execute the instruction on different portions of data in parallel, thus allowing more computations to be performed in a shorter time [65]. This mode of processing fits low-level and intermediate level operations well as they require applying the same operation to different pixel data. Naturally, SIMD can also be used to speed up matrix–vector operations.

The SIMD concept has been used extensively since the 1980s as evident from its widespread use in vision accelerator boards, instruction set extensions for general-purpose processors (GPPs), and packed data processing of digital signal or media processors. In fact, the most common instantiation of the concept of SIMD in today's GPPs, digital signal and media processors, is in the form of the packed data processing extension, which is also known as subword parallelism or wordwide data optimization [32, 65, 81, 124]. These extensions have primarily been developed to help speed up the processing of multimedia data. Since pixel data are usually represented by 8-bits or 16-bits, and since most modern processors have 32-bit registers, packed data processing allows packing four 8-bit pixels or two 16-bit pixels into 32-bit registers and then issuing an instruction to be performed on the individual 8-bit or 16-bit pixels at the same time. These types of packed data instructions not only alleviate the computation burden of low-level and intermediate-level operations, but also help to reduce memory access bottlenecks because multiple pixel data can be read using one instruction. Packed data processing is a basic form of SIMD. In general, SIMD is a useful tool for speeding up low, intermediate, and matrix–vector operations on modern processors. Thus, one can think of SIMD as a tool for exploiting data level parallelism (DLP).

While SIMD can be used for exploiting DLP, VLIW can be used for exploiting instruction level parallelism (ILP) [65], and thus for speeding up high-level operations [20]. VLIW furnishes the ability to execute multiple instructions within one processor clock cycle, all running in parallel, hence allowing software-oriented pipelining of instructions by the programmer. Besides the fact that for VLIW to work properly there must be no dependencies among the data being operated on, the ability to execute more than one instruction per clock cycle is essential for image/video processing applications that require operations in the order of giga operations per second [20].

Of course, while SIMD and VLIW can help speed up the processing of diverse image/ video operations, the time saved through such mechanisms would be completely wasted if there did not exist an efficient way to transfer data throughout the system [35]. Thus, an efficient memory subsystem is considered a crucial component of a real-time image/video processing system, especially for low-level and intermediate-level operations that require massive amounts of data transfer bandwidth as well as high-performance computation power. Concepts such as direct memory access (DMA) and internal versus external memory are important. DMA allows transferring of data within a system without burdening the CPU with data transfers. DMA is a well-known tool for hiding memory access latencies, especially for image data. Efficient use of any available on-chip memory is also critical since such memory can be accessed at a faster rate than external memory. More discussion on memory usage optimization techniques are mentioned in Chapter 4.

One key problem with current memory subsystems is that they were originally designed for one-dimensional data access and thus cannot properly address the spatial locality necessary for two-dimensional or three-dimensional image data. Some researchers have dealt with this problem by designing custom memory addressing schemes that allow for more efficient memory access of image data, as will be seen in the examples section of this chapter. Before stating these examples, it is useful to mention an overview of the standard processor architectures and their advantages/disadvantages for real-time image/video processing.

3.3   OVERVIEW OF CURRENTLY AVAILABLE PROCESSORS

3.3.1   Digital Signal Processors

Digital signal processors are well known for their high-performance, low-power characteristics, and relatively small size, which enable them to accelerate computationally intensive tasks on embedded devices. While it may have been true in the past that digital signal processors (DSPs) were not suitable for processing image/video data in that they could not meet real-time requirements for video rate processing, this is no longer the case with newly available high-performance DSPs that contain specific architectural enhancements addressing the data/computation throughput barrier.

DSPs have been optimized for repetitive computation kernels with special addressing modes for signal processing such as circular or modulo addressing. This helps to accelerate the critical core routines within inner loops of low- and intermediate-level image/video processing operations. In many DSP implementations, it is observed that a large percentage of the execution time is due to a very low percentage of the code, which simply emphasizes the fact that DSPs are best for accelerating critical loops with few branching and control operations, which are best handled by a GPP [1]. DSPs also allow saturated arithmetic operations that are useful in image/video processing to avoid pixel wraparound from a maximum intensity level to a minimum level or vice versa [124]. DSPs possess either a fixed-point or a floating-point CPU, depending on the required accuracy for a given application. In most cases, a fixed-point CPU is more than adequate for the computations involved in image/video processing.

DSPs also have predictable, deterministic execution times that constitute a critical feature for ensuring that real-time deadlines are met. In addition, DSPs have highly parallel architectures with multiple functional units and VLIW/SIMD features, further proving their suitability for image/video processing. DSPs have been designed with high memory bandwidth in mind, on-chip DMA controllers, multilevel caches, buses, and peripherals, allowing efficient movement of data on- and off-chip from memories and other devices. DSPs support the use of real-time operating systems (RTOSs), which again help in guaranteeing that critical system level hard real-time deadlines are met. Of course, DSPs are fully programmable, which adds to their inherent flexibility to changes in algorithm updates. Modern development tools such as efficient C code compilers and use of hardware-specific intrinsic functions have supplanted the need to generate hand-coded assembly for all but the most critical core loops, leading to more efficient development cycles and faster time-to-market.

Indeed, DSPs contain specific architectural features that help one to speed up repetitive, compute-intensive signal processing routines, making them a viable option for inclusion in a real-time image/video processing system. That is why DSPs have been used in many real-time image/video processing systems. More recently, DSPs have been included as a core in dual-core processor system-on-chips for consumer electronics devices such as PDAs, cell phones, digital cameras, portable media players, etc.

3.3.2   Field Programmable Gate Arrays

Field programmable gate arrays (FPGAs) are arrays of reconfigurable complex logic blocks with a network of programmable interconnects [148]. The amount of gates and capabilities of FPGAs are expected to continue to grow in future generations. FPGAs allow fully application-specific custom circuits to be designed by using a software programming language known as hardware description language (HDL). They provide precise execution times helping to meet hard real-time deadlines. FPGAs can be configured to interface with various external devices. Since they are reprogrammable devices, they are flexible in the sense that they can be reconfigured to form a completely different circuit. Current generation FPGAs can be either fully reconfigured or partially reconfigured, with reconfiguration times of less than 1 ms, making it possible to have a dynamic run-time reconfiguration. This configuration is useful for reducing system size of embedded devices. The interested reader can refer to [2, 14, 22, 83, 140] for more information on run-time reconfigurations of FPGAs for image/video processing applications.

Due to their programmable nature, FPGAs can be programmed to exploit different types of parallelism inherent in an image/video processing algorithm. This in turn leads to highly efficient real-time image/video processing for low-level, intermediate-level, or high-level operations, enabling an entire imaging system to be implemented on a single FPGA. In general, FPGAs have extremely high memory bandwidth. As a result, one can use custom memory configurations and/or addressing techniques to exploit data locality in high-dimensional data. In many cases, FPGAs have the potential to meet or exceed the performance of a single DSP or multiple DSPs. FPGAs can be thought of as combining the flexibility of software programmability with the speed of an application-specific circuit (ASIC) within a shorter design cycle or time-to-market. Often an FPGA implementation is the first step toward transitioning to an ASIC, or in some cases the final product. However, there is a disadvantage associated with FPGAs, that is, their energy or power consumption efficiency. Lately, low-power FPGAs are becoming more available.

In essence, FPGAs have high computational and memory bandwidth capabilities that are essential to real-time image/video processing systems. Because of such features, there has been an increasing interest in using FPGAs to solve real-time image/video processing problems [38]. FPGAs have already been used to solve many practical real-world, real-time image/video processing problems, from a preprocessing component to the entire processing chain. FPGAs have also been used in conjunction with DSPs. A current trend in FPGAs is to include a GPP core on the same chip as the FPGA for a customizable system-on-chip (SoC) solution.

3.3.3   Multicore Embedded System-on-Chip

In the consumer electronics market, there has been a drive toward single-chip solutions or SoCs for portable embedded devices, which require high performance computation and memory throughput coupled with low power consumption in order to meet the real-time image/video processing constraints of battery-powered products such as digital cameras, digital video camcorders, cell-phone-equipped cameras, etc. These systems exhibit elegant designs where one can learn how the industry has approached the battery-powered embedded real-time image/video processing problem.

For example, consider the TMS320DM320 "digital media processor" manufactured by Texas Instruments [159]. This is a multiprocessor chip with a reduced instruction set (RISC) microprocessor coupled with a low-power fixed-point DSP. The RISC microprocessor serves as the master handling system control, running a RTOS and providing the necessary processing power for complex control-intensive operations. The DSP, acting as a slave to the RISC, is a low-power component for performing computationally intensive signal processing operations. The presence of a memory traffic controller allows achieving a high-throughput access to memory. In this device, the RISC and DSP are accompanied by a set of parameter customizable application-specific processors that provide a "boost," that is to say, they provide the extra computational horsepower that is necessary to perform functions such as real-time LCD preview (Preview Engine) and real-time computation of low-level statistics necessary for autoexposure, autowhite balance, and autofocus (H3A Engine). The DSP along with its accelerators and dedicated image processing memory buffers provides a high-computation throughput and memory bandwidth for performing various image/video processing related functions such as rendering the final captured image through the image pipeline and running image/video compression routines.

By examining this architecture, one can see that this SoC has been designed with a DSP plus dedicated hardware accelerators for low-level and intermediate-level operations along with a GPP hardware for more complex high-level operations. This is an illustrative example showing that a complete real-time image/video processing system can be characterized as a heterogeneous architecture with a computation-oriented front end coupled with a general-purpose processing back end. Of course, the TMS320DM320 is just one good example of many currently available multiprocessor embedded SoCs. In fact, as it will be seen in the examples section, the low-power, moderate performance DSPs plus accelerators have been widely used by many research groups in the form of DSP/FPGA hybrid systems, most likely due to cost issues associated with ASIC development.

An interesting recent hardware development for digital imaging is the Texas Instrument DaVinci technology that couples an Advanced RISC Machines (ARM) processor with a high-performance C64x DSP core [160]. This technology provides the necessary processing and memory bandwidth to achieve a complete imaging SoC. Examples of research performed on multicore embedded SoC for digital camera applications can be found in the references [52, 78, 79, 108, 115, 116], which cover the development and implementations of the automatic white balancing, automatic focusing, and zoom tracking algorithms encountered in today's digital camera systems.

3.3.4   General-Purpose Processors

There are two types of GPPs on the market today, one geared toward nonembedded applications such as desktop PCs and the other geared toward embedded applications. Today's desktop GPPs are extremely high-performance processors with highly parallel architectures, containing features that help to exploit ILP in control-intensive, high-level image/video operations. SIMD extensions have also been incorporated in their instruction sets allowing such processors to exploit DLP and enabling moderate acceleration of multimedia operations corresponding to low-level and intermediate-level image/video processing operations. GPPs have been outfitted with the multilevel cache feature. This feature provides the potential of having low latency memory accesses for frequently used data. These processors also require an RTOS in order to guarantee a real-time execution. Desktop GPPs are characterized by their large size, requiring a separate chip set for proper operation and communication with external memory and peripherals.

Although GPPs have massive general-purpose processing power, they are extremely high-powered devices requiring 100s of watts of power. Clearly such processors are not suitable for embedded applications. Despite this fact, advances in desktop GPPs have allowed the standard commercial off-the-shelf desktop PCs to be used for implementing nonembedded real-time image/video processing systems. In [100], it is even claimed that the desktop PC is the de facto standard for industrial machine vision applications where there is usually enough space and power available to handle a workstation. It should be noted that such industrial inspection systems usually augment the processing power of the desktop GPP with vision accelerator boards. These boards often furnish a dedicated SIMD image/video processor for high-performance real-time processing not normally met by the SIMD extensions to the desktop GPP. Recently, a paradigm shift toward multicore processor designs for desktop PCs has occurred in order to continue making gains in processor performance.

On the embedded front, there are also several GPPs available on the market today with high-performance general-purpose processing capability suitable for exploiting ILP coupled with low power consumption and SIMD-type extensions for moderately accelerating multimedia operations, enabling the exploitation of DLP for low-level and intermediate-level image/video processing operations. Embedded GPPs have been used in multicore embedded SoCs, providing the horsepower to cope with control- and branch-intensive instructions. Both embedded and desktop GPPs are supported by mature development tools and efficient compilers, allowing quick development cycles. While GPPs are quite powerful, they are neither created nor specialized to accelerate massively data parallel computations.

3.3.5 Graphics Processing Unit

The early 2000s witnessed the introduction of a new type of processor, the graphics processing unit (GPU). The primary function of such processors is for real-time rendering of three-dimensional (3D) computer graphics enabling fast frame rates and higher levels of realism required for state-of-the-art 3D graphics in modern computer games. While the original GPUs were fixed function accelerators, current generation GPUs incorporate more flexibility through ever-increasing amounts of programmability with programmable vertex and texture/fragment units that are useful for customizing the rendering of 3D computer graphics. GPUs can also be used for accelerating computations with inherent DLP. In terms of performance, for example, an Intel 3.0-GHz Pentium 4 GPP provides 12 GFLOPS peak floating-point computational performance and 5.96-GB/s memory throughput, while the ATI Radeon X1800XT GPU provides 120 GFLOPS peak floating-point performance with 42-GB/s memory throughput [64]. This shows that GPUs can provide huge increases in GFLOPS performance and memory throughput over those of a high-performance desktop GPP.

Due to their floating-point calculation capabilities, the increased levels of programmability, and the fact that GPUs can be found in almost every desktop PC today, many researchers have been looking into ways to exploit GPUs for applications other than the real-time rendering of 3D computer graphics, an area of research referred to as general-purpose processing on the graphics processing unit (GPGPU). GPUs have already been deployed to solve real-time image/video processing problems including complete computer vision systems [21, 50], medical image reconstruction in magnetic resonance imaging (MRI) and ultrasonic imaging requiring FFT [136], stereo depth map computation [153], and subpixel accurate motion estimation at video rates [82]. A recent survey paper on the state-of-the-art in GPGPU [110] also presents several examples of how the power of GPUs has been applied to calculation-intensive problems in signal and image processing including identifying a 3D surface embedded in an MRI volume image, which is considered a difficult medical image segmentation problem, image registration, real-time simultaneous computation and visualization of motion, and tomography reconstruction.

To understand how a GPU can be used to perform image processing, let us take a look at its graphics pipeline [82]. The pipeline begins with a set of vertices or points in a 3D space to define graphics primitives, followed by applying vertex shaders to allow programmable control over vertex transformations. Once the graphics primitives have been defined, textures are then mapped onto them to add scene details. Texture shaders, known as fragment programs or kernel fragments, can also be applied to the textures. In the final stage, the pixels are rendered to either a display frame buffer or an offscreen rendering buffer known as pixel buffer for further texture processing through the graphics pipeline. Thus, image processing on a GPU can be performed by downloading an image to the GPU as a texture structure, rendering a rectangle the size of the image, and mapping the image as a texture structure to the rectangle, after which a kernel fragment program can be used to process the image taking advantage of the massive computation power of the GPU.

One key drawback of GPUs has been the data read-back throughput through the Peripheral Component Interconnect (PCI) bus, but this is expected to be mitigated with the introduction of the PCI Express bus standard. One important item to note is that just like desktop GPPs, GPUs are also high-powered devices drawing 100s of watts of power. Although low-power GPUs for embedded applications are becoming more available, it is currently not known how well these embedded GPUs will fare in GPGPU applications. The reader is referred to [158] for more information on GPGPU.

3.4   EXAMPLE SYSTEMS

3.4.1   DSP-Based Systems

Due to their high computation performance levels coupled with their low power consumption, DSPs have been used extensively in embedded devices for accelerating computation heavy components of an image/video processing algorithm in many applications such as image filtering, video surveillance, and object recognition.

3.4.1.1   Image Filtering Operations

Image filtering operations are well suited for implementation on DSP platforms as their regular, repetitive looping computation structure fits the DSP architecture well. Because of this, many attempts have been made in implementing image filtering operations on DSPs. One of the more computationally challenging filtering problems involves the use of nonlinear filters. Such filters are often employed for the purpose of removing impulsive-type noise while at the same time maintaining the integrity of edges. One research group has consistently shown the usefulness of a single-chip, high-performance DSP for performing real-time nonlinear image filtering. Several examples of their implementation of nonlinear filtering algorithms can be found in [51, 117–120]. Due to the high computational complexity of the algorithms, high-performance, floating-point VLIW DSPs were chosen for the implementations. While the TMS320C6701 DSP running at 167 MHz was used in [51, 117, 118], the TMS320C6711 DSP running at 150 MHz was used in [119, 120]. In [51, 118, 119], it was shown that by using the DSP platform, a real-time video rate edge-preserving, nonlinear filtering could be achieved for Quarter Common Image Format (QCIF) sized video sequences. In [120], the extension of the nonlinear filtering algorithms to 3D was demonstrated using a single-chip DSP platform.

3.4.1.2 Computationally Complex Operations

Another computationally complex algorithm for which a single-chip, high-performance DSP has been utilized is automatic color reduction in the CbCr chrominance color space using the two-dimensional (2D) version of a multiscale clustering algorithm [107]. For this application, the fixed-point VLIW TMS320C6201 DSP running at 133 MHz was used as the implementation platform. The performance achieved was 20 s for 256×256 images, showing just how computationally demanding an algorithm could be even on a high-performance DSP.

3.4.1.3 Entire Image Processing Chains

In contrast to single-chip, high-performance DSP solutions, multichip, low-performance DSP solutions have also been popular implementation platforms, especially for low-cost implementation of a complete image/video processing chain. One example of such a chain can be seen in [69], where a video surveillance system for detecting people in complex outdoor environments subject to lighting and background changes was implemented using a total of nine, low-performance TMS320C40 DSPs connected via their data bus and communication ports. This setup was able to achieve a satisfactory real-time performance of 15 fps for $1024 \times 256 \times 32$-bit video.

Depending on the complexity of the algorithm involved, sometimes a single-chip DSP platform is adequate for implementing a complete system. A good example of such a case can be seen in [98], where the problem of real-time recognition by a small autonomous soccer playing robot was presented. Due to the dynamic nature of the environment in which the system was to operate, a robust operation against changing lighting conditions and partial occlusions at 60 fps was required. To meet these requirements, a single-chip, high-performance DSP platform was chosen over an FPGA platform in order to have a lower power consumption, an easier development, and a lower total cost. The Analog Devices ADSP-BF533 Blackfin DSP was chosen furnishing 1200-M multiply-and-accumulate (MAC) operations at 600 MHz and consuming 280-mW power. This met the requirements of having a low power consumption of less than 500 mW and an estimated performance of 800 million MAC operations.

From the above examples, it can be seen that both a single high-performance DSP and multiple low-performance DSPs have been used to implement image/video processing algorithms. As mentioned earlier, only recently DSPs have been equipped with the ability to process image/video data at video rates [160]. It is expected that the use of DSPs will continue to grow well into the future.

3.4.2   FPGA-Based Systems

Due to their flexibility in implementing custom hardware solutions, FPGAs have been used extensively for implementing a single component of an image or video processing system all the way up to the entire system. The main reason often cited for using FPGAs over other platforms is that they provide a low-cost, flexible development of high-performance, custom parallel processors, suitable for transitioning almost any kind of image/video processing algorithm from a development environment to a real-time implementation.

3.4.2.1 Image Filtering Operations

FPGAs have been used to implement various types of image filtering problems. An example of 2D nonlinear image filtering implemented on a field programmable logic device (FPLD) appears in [131], where the problem of mammogram contrast enhancement in real-time was addressed. The slow execution time in software motivated the desire to search for a hardware solution. The FPLD was chosen as the implementation platform for its ability to achieve higher processing performances as compared to GPPs and DSPs, and for its flexible development characteristics. The reported results showed that this implementation allowed the filtering to be performed within 98 ms with an 8-MHz clock and 23 ms with a 33.3-MHz clock for $512 \times 512 \times$ 8-bit mammograms. Due to the flexibility of the FPLD, the modifications needed to have 12-bit accuracy were easy to incorporate into the system, only requiring making changes to the filter data-path. Another example of 2D image filtering can be found in [59], where the real-time feasibility of using fuzzy morphological filters for processing image/video sequences was discussed. An implementation of these filters was performed using the Xilinx Virtex XCV300

FPGA achieving a performance of 179 fps for 512 × 512 images. The reported results showed that the fuzzy morphological filters outperformed other filters.

Recently, 3D image filtering operations have been implemented on FPGA platforms. In these implementations, custom memory access schemes combined with high-performance memory subsystems have enabled the real-time throughput essential for 3D image processing. One example of such an implementation can be found in [24], where real-time anisotropic diffusion filtering was performed on 3D ultrasonic images for the removal of speckle noise. Due to the complexity of anisotropic diffusion filtering and the large amount of data in 3D images, the software implementation could not generate a real-time throughput. A hardware implementation based on a single FPGA was chosen to meet the real-time requirement. The key aspects of the architecture included a custom, efficient 3D image data access scheme called "brick buffering," which allowed an optimized, high-throughput access to 3D image data. The implementation was performed on an Altera Stratix II EP260F484C3ES FPGA running at a 200-MHz clock rate with two parallel 100-MHz, 32-bit external SDRAMs for input/output. A real-time performance of 24 iterations per second for 128 × 128 × 128 images was achieved.

Another example showing the power of an FPGA for accelerating 3D image preprocessing tasks can be found in [145], where the problem of 3D median and convolution filtering for 3D medical image processing was considered. Since software implementations could not meet the real-time constraints, a hardware-based solution using an FPGA was considered. The FPGA platform was chosen due to its high-performance capabilities and its flexibility. This platform made it possible to implement the median filtering and convolution operations using fast multipliers. The implementation was performed on the Xilinx Virtex II Pro 2VP125FF1696-6 FPGA achieving a performance of 95 fps for 128 × 128 × 128 images and 12 fps for 256 × 256 × 256 images.

3.4.2.2 Low-Level Operations

Normally, a single-chip FPGA solution is used specifically for accelerating low-level operations, passing on the results for high-level interpretation operations to a GPP. For example, in [13], the problem of controlling the exposure time of a charge-coupled device camera in real-time was considered using a low-level histogram-based measure. Due to the large amount of data that needed to be processed, a hardware-based solution was deemed necessary to meet the real-time requirement. The developed solution was a combination of an FPGA and software running on a host PC. The FPGA was used for the histogram calculation and noise-level calculation, the results of which were then sent to the GPP for further processing.

Another example exhibiting the use of an FPGA for accelerating low-level operations can be seen in [53], where the problem of estimating position and velocity of objects by a three-camera stereo vision system was presented using area-based features. A key requirement of the system was that it had to process images at video rates, posing the need for a dedicated hardware to meet the real-time requirement. The FPGA technology was chosen as the hardware part of a software/hardware-based solution. The tracking problem was broken up into several subtasks including segmentation, correspondence, and motion estimation. The low-level image processing operations of segmentation, noise filtering, and area measurement were implemented on the FPGA, while the higher level operations consisting of extended Kalman filtering and prediction were implemented in software. The utilized FPGA was Altera Flex 10-K100, which met the real-time requirement of the PAL signal video rate.

3.4.2.3 Standard Image Processing Operations

FPGAs have also been used for implementing basic, but computationally expensive tasks encountered in many image/video processing applications such as edge detection, moment calculation, and Hough transform.

Regarding edge detection, in [66], a subpixel edge detection algorithm for an industrial inspection application was presented. A single FPGA platform was chosen as the desired implementation vehicle over analog processing, a custom VLSI solution, and a hybrid FPGA/DSP solution. A single FPGA, combined with a computationally simple algorithm, provided the required detection rate and reduction in the computation and system cost. A Xilinx XC-4005E FPGA was used along with a Xilinx Foundation ISE to synthesize and implement the developed VHDL code. The implementation was able to process high-resolution 1024 line scan images at 2000 fps using a 200-MHz clock.

Geometric moments are used extensively as key image features in many image/video processing applications, but due to the computational complexity involved in their calculation, their real-time implementation cannot be easily achieved. The problem of computing geometric moments in real time was considered in [88]. Since a real-time performance could not be achieved by using standard processors, an FPGA solution was considered. The developed algorithm was implemented on an Altera EP1K50TC144-1 FPGA, a member of the ACEX1K FPGA family, using the MAX+PLUS II environment. The results showed that ten moments of a 1 megapixel image could be processed within 25 ms.

FPGAs have also been used to accelerate the computationally complex Hough transform, which forms the basis of many image/video processing algorithms. In [39], an efficient Hough transform implementation was presented. Noting that, in general, the Hough transform implementation requires a great deal of computation horsepower, a hardware approach was desired and thus an FPGA implementation was chosen over an ASIC one for its implementation flexibility. The Hough transform algorithm was also simplified before the implementation on the FPGA by reducing the use of lookup tables and increasing the parallelism of the calculations. The Virtex II Xilinx XC 250-5FG456C FPGA was chosen, which at the clock rate of 606 MHz generated four line values in parallel every 12 ns.

3.4.2.4 Compression Operations

When dealing with the compression of a huge amount of data such as HDTV data, FPGAs can help to cope with the massive data throughput, enabling an effective means of achieving real-time compression.

For example, in [144], the problem of computing the 2D discrete, biorthogonal wavelet transform (DBWT) for HDTV video compression was discussed. It was noted that the computation of the DBWT was the most time-consuming part of the video compression algorithm and that software implementations were not able to meet the real-time requirement, thus requiring a hardware-based solution. An FPGA was chosen as an alternative to an ASIC solution and for its flexible architecture that allowed it to be quickly reconfigured using user-defined adjustable compression parameters. The implementation and verification were performed on the Celoxica RC1000-PP PCI-based FPGA development board containing a Xilinx Virtex 2000E FPGA. The implementation was able to achieve a real-time performance of 286, 139, and 121 fps, respectively, for the three wavelet decomposition levels on 1280 × 720 resolution input images and 127, 62, and 54 fps, respectively, for the three wavelet decomposition levels on 1080 × 1920 resolution input images.

Another example of a real-time compression application enabled with the use of FPGA can be seen in [106], where the problem of real-time image compression for high-speed cameras was addressed. High-speed cameras are characterized by extremely high frame rates of the order of thousands of frames per second. Standard image compression methods are just not capable of keeping up with the high input data rate of these cameras, and thus the need for a hardware solution. A compression engine was proposed using 32 parallel image compression circuits. In total, seven FPGAs were employed in the design, one for an input buffer, one for an output buffer, one for a master controller, and four for implementing the 32 parallel image processing circuits. A frame rate of 2000 fps was achieved for 512 × 512 images.

3.4.2.5 Entire Image Processing Chains

Entire image processing systems have also been implemented using FPGAs, including face detection and tracking, inspection, stereo vision, and 3D image registration systems.

Recently, much research has been done on practical systems for face detection and tracking. One such system has been developed and shown to work in real-time on an FPGA device in [112], where face detection and tracking was considered in the resource constrained environment of an embedded mobile device. It was stated that many methods for face detection were too resource demanding for an embedded device, requiring extremely high-performance computation power, large amounts of memory, and floating-point operations. A simplified algorithm was developed and implemented on an Altera EP20K1000EBC652-1 FPGA. The FPGA was chosen as the implementation vehicle for rapid prototyping, enabling a proof of concept test before a full VLSI implementation. The achieved performance was 434 fps at a clock rate of 33 MHz.

Inspection systems often require a high-performance processing subsystem in order to cope with the involved hard real-time constraints. An FPGA is well suited for such applications. As discussed in [67], a low-cost, high-performance system for checking multiple-choice question sheets was developed by using a high-speed optical mark reader (OMR). OMRs are used for processing large amounts of data in a relatively short amount of time. However, the cost of such systems is often excessively high, limiting their widespread use. Thus, it was desired to implement the OMR system on a single FPGA to lower the system cost while having a customizable parallel processing capability. The system was implemented on a Xilinx SpartanIIE XC2S300E FPGA using VHDL, and the implementation was able to process the 3456 pixel line sensor images at a rate of 5000 fps with a 20-MHz clock. The real-time operation at 5000 fps eliminated the need for large memory storage, thus reducing the system cost.

Another computationally intensive operation involves computing the depth information of a scene utilizing stereo image processing techniques. As shown in [7], an FPGA was used to achieve a real-time implementation of a dense disparity map computation using a correlation-based approach. This implementation was designed to minimize external memory accesses and to perform parallel processing of different correlation windows. The utilized FPGA belonged to the XCV800HQ240-6 Virtex family from Xilinx, and was able to produce a real-time performance of 60 disparity computations per second for 320×240 images.

In medical imaging, practical deployment of computationally demanding 3D image processing is of much interest. In [23], the calculation-intensive problem of 3D multimodality image registration, which is essential for practical deployment of image-guided medical procedures, was considered. Past solutions to this problem involved using a supercomputer implementation, which was not practical in a hospital setting. Thus, a custom hardware solution based on an FPGA was used, achieving a speedup comparable to a 64 parallel processor supercomputer. The utilization of parallel memory accesses and a parallel calculation pipeline was the key to obtaining a considerable amount of speedup. An FPGA running at 200 MHz with 100-MHz memory buses was used in conjunction with high-speed SDRAM and SRAMs. In addition, the required lookup table was implemented in one 512-k memory block and all the calculations were implemented using 32-bit fixed-point numbers. The developed architecture was able to process 50 million voxels per second, providing the real-time throughput necessary in a practically useful image-guided medical treatment.

As one can see from the above diverse examples, FPGAs have been extensively deployed as flexible, custom processors for solving real-time image/video processing problems primarily due to their ability to exploit different types of parallelism inherent in an image/video processing algorithm. However, it must be noted that, in general, most of these FPGA solutions are meant to be accelerators that are hosted by a PC and not for embedded devices.

3.4.3   Hybrid Systems

There have also been many examples in the literature regarding hybrid systems, which include some combination of DSP and FPGA processors. In these systems, an FPGA is often used as a preprocessor performing the function of a parallel pixel processor for low-level and intermediate level operations, while a DSP is used for handling intermediate and high-level operations or other computationally simpler matrix–vector operations. Such systems have been shown to be capable of supporting the real-time demands of an entire image/video processing chain. Eight system examples are mentioned below to further illustrate the usefulness of a hybrid solution.

3.4.3.1 Image Segmentation Systems

An image segmentation system consists of a diverse set of operations. To meet the real-time requirements of an image segmentation system, a hybrid FPGA and DSP solution was used in [3] to implement the diverse set of operations involved. These operations included leveling, regularization, and reconstruction. The low-level operations of minimum extraction and lower/upper regularization were performed on an FPGA, while a TMS320C44 DSP was used for the implementation of the high-level operations of reconstruction and fusion of minima.

Another hybrid platform for image segmentation was reported in [4] and [5], where the segmentation problem based on thinning and crest restoration was considered. The only difference between the two references was a change in hardware, where in [4] a Mirotech Arix board with a Xilinx Virtex XCV300 FPGA and a TMS320C44 DSP were used, while in [5] an XA10 Excalibur SoC with a 32-bit RISC ARM922T microprocessor core and an Apex 20KE PLD were used. In [4], the entire segmentation chain was implemented on the FPGA, which allowed processing of 512 × 512 images at 125 fps. In [5], the crest restoration was implemented in software running on the ARM processor to provide more flexibility in the implementation, but at the expense of a considerable reduction in performance, causing the entire chain to take 6 s to process one 512 × 512 image.

3.4.3.2 Industrial Inspection Systems

Industrial inspection systems are characterized by a diverse processing chain. For example, in [135], an automatic quality control system for textile fabrics was designed. The system was implemented by using an FPGA for the synchronization, a DSP for the high-level texture feature extraction and neural network classification, and a host PC for the defect detection and geometric feature extraction.

Another inspection application that was successfully implemented using a hybrid platform was reported in [125], where the problem of locating the intersection of horizontal and vertical crossbars with subpixel accuracy was addressed. The real-time requirement of the embedded system was getting subpixel accurate location detection on 1024 × 1024 images at a rate of 50 fps, leaving an upper bound of 20 ms for the processing. To meet this requirement, a hybrid architecture was employed using three Xilinx XC4000 FPGA for high data rate, low-level operations (area location, horizontal and vertical center of mass calculations), and two TMS320C44 DSPs for low data rate, high-level operations (linear regression for line calculation). The FPGAs were chosen for their fast arithmetic and internal RAMs and ROMs, while the DSPs were chosen for their four communication ports, floating-point arithmetic, DMA coprocessor, memory buses, and 2k-word RAM and cache. To synchronize the communication between the DSPs and the FPGAs, the DSP's communication protocol was implemented on the FPGAs.

3.4.3.3 Video Compression Systems

Video encoding systems utilizing wavelet transform coding techniques can also benefit from a hybrid platform. Such an approach was considered in [31], where the entire Motion-JPEG2000 video encoder was implemented on the high-performance TMS320C6416 VLIW DSP, achieving encoding speeds at full video rates of 30 fps. In this application, an FPGA was used for merging the digitized image data fields into one frame, for transferring of image data, and for providing overall system control, while the encoding was implemented entirely on the DSP.

3.4.3.4 Smart Camera Systems

Emerging smart cameras consist of a diverse set of image/video processing algorithms and are often implemented using hybrid platforms. One example of such platforms can be seen in [17], where a smart digital video camera surveillance system was introduced consisting of a combination of an FPGA and a multiprocessor DSP configuration. Two TMS320C6415T DSPs were chosen, each providing up to 8000 MIPS at 1 GHz and 1 MB of internal memory. The large amount of internal memory was necessary to achieve an efficient implementation. The FPGA provided the necessary glue logic for interfacing the DSP units to the image sensors. Multiple DSPs were used since no single-chip DSP solution existed to satisfy the processing requirements, which included processing of 720×576 color video streams as opposed to small-resolution CIF and QCIF images commonly utilized in camera-equipped cell phones.

Similar to smart cameras, the image processing chain of autonomous navigation systems is also a good candidate for the utilization of a hybrid platform. For example, in [10], the problem of real-time underwater imaging for autonomous vehicle navigation at video rates was discussed. To address the needs for such a system, a 2D array of FPGA and DSP processors was constructed for pipelined, parallel processing. Each processing element consisted of a ping-pong style memory buffer, a TMS320C51 DSP for computation, and an FPGA for communication and low-level image processing operations. For this application, the FPGA performed the image processing tasks, while the DSP computed the angular displacement and distance parameters.

These examples have illustrated that various combinations of FGPAs and DSPs can be used to solve real-time image/video processing problems. In such systems, an FPGA usually performs low-level to intermediate level operations, while a DSP handles intermediate to some computation-oriented, high-level operations. In general, hybrid platforms are suitable for real-time implementation of those image/video processing chains that incorporate a diverse set of operations.

3.4.4   GPU-Based Systems

GPU-based developments in the field of real-time image/video processing are fairly new. Therefore, only two examples from the literature are presented here including stereo depth map computation and subpixel motion estimation.

3.4.4.1 Stereo Vision System

The problem of computing a complete depth map for a stereo vision system was considered in [153]. It was pointed out that while the real-time calculation of stereo depth maps was possible with standard desktop GPPs, primarily due to advancements in clock speeds and SIMD instruction set extensions, such an implementation taxed the system to the extent that there were no more resources left to perform high-level, control-intensive, interpretation tasks. It was thus decided to make use of the computation power of the GPU for off-loading the depth map computation. This freed up resources to execute the high-level interpretation operations on the GPP. The power of the GPU also allowed the use of advanced features, including multiresolution matching, adaptive windowing, and cross-checking, not found in standard implementations. The results indicated that 289 million disparity evaluations per second could be achieved on the ATI Radeon 9800 GPU for 512 × 512 images and a 94-pixel disparity range.

3.4.4.2 Motion Estimation System

The problem of subpixel accurate motion estimation for improving the quality and efficiency of the standard video compression schemes was considered in [82]. It was pointed out that most of the standards for video coding recommend using subpixel accurate motion estimation for the highest quality, the only caveat being that the interpolation operation presents a huge computational burden. The performance goal of the system was to perform subpixel accurate motion estimation for 720 × 576 images at 25 fps using the full search algorithm, which was not feasible without some hardware assistance. To meet this real-time requirement, a GPU was used to perform the interpolation and the block matching motion estimation algorithm. The interpolation was performed using the GPU's inbuilt bilinear interpolation function and the motion estimation algorithm was restructured to make a better use of the available resources on the GPU. In all, a four times speedup over a GPP implementation was achieved, the primary bottleneck being the data read-back bandwidth between the GPU and the PC over an Advanced Graphics Port (AGP) bus. It was stated that better performance gains could be achieved with the newer PCI Express bus.

 As observed from these two examples, GPUs have the potential to solve computationally intensive, data parallel real-time image/video processing problems. The standard use of a GPU is to accelerate computationally intensive operations, leaving the GPP of its host free to handle other tasks. With GPU performance growing at an ever-increasing rate and the introduction of faster bus architectures, such as the PCI Express, the popularity of using GPUs for solving real-time image/video processing problems is expected to increase.

3.4.5   PC-Based Systems

PC-based systems have also been widely used for solving real-time image/video processing problems. Such systems are usually equipped with a camera and a frame grabber, using the PC as a host. Four examples of such systems are mentioned next.

3.4.5.1 Object Detection System

Object detection is a computationally complex problem, requiring a high-performance processor for practical implementations. In [152], the problem of object detection in real-time was discussed. A point was made that while VLSI, ASIC, or FPGAs can be used to meet the real-time constraint for video rate object detection, such solutions require a low-level hardware design that is often difficult to achieve by image processing developers unfamiliar with design techniques. Thus, it was decided to use the Datacube MaxPCI vision accelerator board that provided the necessary parallel computation power and high data throughput to process 1000×1000 images at 30 fps.

3.4.5.2 Computer Vision System

A computer vision system involves many diverse operations that map well to vision accelerator boards. For example, in [100], a generalized, scalable and modular architecture for a real-time computer vision application based on desktop PCs was presented. The architecture consisted of an image acquisition module and a PC-based processing module, where both modules could be scaled to handle more cameras and higher processing demands. The system was applied to an industrial inspection application involving quality control of TV screen manufacturing. The implemented system made use of eight JAI CV-M10BX CCIR cameras and four Matrox Meteor II/MC frame grabbers with the PCs equipped with dual Pentium III processors running at 600 MHz.

3.4.5.3 Video Segmentation System

Another computationally complex problem involves real-time segmentation of video data. It has been shown in [154] that such a system can be implemented using off-the-shelf components without the need for high-end and expensive frame grabbers. In this reference, the problem of image sequence segmentation based on a global camera motion compensation, a robust frame differencing, and a curve evolution was discussed. A computationally efficient algorithm was developed and implemented for use on a PC with a Pentium I 400-MHz processor. Video acquisition was done using a 3Com Home Connect USB Web Camera, which eliminated the need for a relatively expensive frame grabber. Of course, this was not meant to be a suitable replacement for higher resolution and higher frame rates. The segmentation performance achieved was 5 fps for 160 × 120 images, keeping in mind that the implementation was done on a rather slow GPP.

3.4.5.4 Image Fusion System

Another example involving the successful use of a vision accelerator board is reported in [132], where an adaptive image fusion algorithm was implemented to aid helicopter pilots. The real-time requirement of processing 256 × 256 images at 25 fps for image registration and a three-level pyramid decomposition was met using a hybrid hardware and software approach. The system consisted of two cameras, each connected to its own Datacube MaxPCI vision accelerator for preprocessing, a 96-MB buffer for storing images from the frame grabber, and a separate accelerator card for image registration.

As revealed from these examples, standard desktop PCs equipped with frame grabbers can be used to solve real-time image/video processing problems. Due to their large size and high power consumption, however, such systems are usually used in industrial inspection settings or those applications where size and power consumption are not critical design issues.

3.5   REVOLUTIONARY TECHNOLOGIES

Around the late 1990s, a fundamentally different approach to real-time image/video processing systems was being developed, namely the idea of fusing the image sensor with the necessary circuitry required for image processing. This was made possible through Complementary Metal-Oxide Semiconductor (CMOS) imaging technology that allows image processing circuitry to be placed on the same die as the image sensor. One of the recent developments along this line is the SIMD pixel (SIMPil) processor [54], which is considered to be a portable multimedia supercomputer, combining the high-performance requirement of multimedia applications with the low power consumption demanded by embedded devices.

The SIMPil processor was used to implement the image processing pipeline found in digital cameras. The simulation results showed that the processing for the entire pipeline for a 1-megapixel Bayer pattern image could be executed in 1 ms on a 500-MHz SIMPil array processor, requiring only 2.8-W power consumption. In addition to this, the utilized SIMPil processor configuration had an estimated peak operation throughput on the order of 1.5 tera operations per second. Given that current digital cameras use a simplified hardwired image pipeline or a preview engine operating at a lower resolution than the full sensor resolution to allow for real-time preview on an LCD, the SIMPil processor could easily do away with the need for a preview engine, and thus allow higher resolution LCD previews. With such an on-sensor-chip image processing, the need for large image memory buffers is eliminated. This leads to a lower system cost. Of course, such a chip would need to be paired with a GPP for high-level operations and system-level control in order to be a complete system.

Technologies like the above impart a radical change in design and performance from current technologies, possessing the capability to usher in a new age for achieving real-time image/video processing.

3.6   SUMMARY

In this chapter, many topics were covered, including key architectural features such as SIMD and VLIW, an overview of DSP, FPGA, multiprocessor SoC, GPP, and GPU platforms, representative example systems from the literature, and future technologies.

It should be noted that each real-time image/video processing application has its own unique needs and requirements including speed, memory bandwidth, power consumption, cost, size, development tools, etc. [95]. Thus, to go from research to reality, it is important to first understand the needs of the system of interest and then pair them up with the appropriate technologies.

CHAPTER 4

Software Methods for Real-Time Image and Video Processing

4.1 INTRODUCTION

Software methods make up another aspect of transitioning an image/video processing algorithm from a research development environment to a real-time environment running on a target hardware platform. More often than not, an algorithm that generates an acceptable performance within a development environment will not, in general, generate the same acceptable performance on the target hardware platform without any modifications. This is especially true with portable embedded devices that are resource limited with generally lower clock rates, limited amounts of memory, and lower power consumption than say a modern desktop PC with a high-performance general-purpose processor (GPP). Because of such limitations, the algorithm must be carefully and properly modified to fit the structure of the underlying hardware, making efficient and effective use of available resources. Items of interest include software architecture design, memory management, and software optimization. While Chapters 2 and 3, respectively, addressed the algorithmic and hardware platform options for implementing real-time image/video processing algorithms, this chapter covers the equally important software side of their real-time implementations. Here, software means the interface between algorithms and hardware. A crucial aspect of having a real-time image/video processing system is the development of efficient software that maximizes the resources associated with the available hardware.

4.2 ELEMENTS OF SOFTWARE PLATFORM

Just as there are many components that make up a hardware platform for real-time image/video processing, there are also many components that make up a software platform. Programming languages, software architecture design principles, and real-time operating systems are some of the key components, which are discussed next.

4.2.1   Programming Languages

For designing image/video processing systems, two types of programming languages are employed, those used for rapid prototyping and development and those used for actual deployment in a stand-alone product. One of the difficulties of software design for real-time image/video processing systems lies in transitioning a source code from the programming language used for development, such as MATLAB® or LabVIEW™, to a source code used for deployment, such as general-purpose C/C++. This section briefly covers the real-time implementation aspects of different programming languages.

4.2.1.1   Research Environment Programming Languages

The programming languages and programming styles used for production-level real-time image/video processing systems are usually quite different from those used merely for research and development during the prototyping design phase. The most common development languages used in the research phase include some combination of MATLAB, LabVIEW, or C/C++.

MATLAB is an interpreted, high-level programming language as opposed to a compiled one. This means that in a MATLAB ".m" source file, each instruction, be it a command or a variable assignment, has to be interpreted each time it is encountered during the execution of the source code. Another programming-related feature of MATLAB is that it is not a strongly typed language, meaning that variables can be declared without any data-type specification since such specifications are interpreted during run-time. Of course, the main attribute of MATLAB lies in its easily accessible matrix–vector processing capabilities as it can handle linear algebra operations on matrix data structures without explicitly using loop constructs. MATLAB can even be outfitted with many function libraries for virtually any type of processing via various toolboxes. Some of the relevant toolboxes include the Image Processing Toolbox, Signal Processing Toolbox, and Fixed-Point Toolbox.

MATLAB is mostly a text-based programming environment using scripts and functions written in ".m" files, but with the Simulink® add-on, it can be turned into a model-based programming environment for graphical-oriented, block-based programming. Simulink, combined with the Image and Video Processing Blockset, further eases the development of an entire image or video processing system as compared to traditional text-based programming by promoting hierarchical, modular design via functional blocks that pass data between them through wires. Simulink also has built-in support for fixed-point data-types, which allows one to explore the numerical aspects of an entire system. Simulink can also be used for hardware-in-the-loop type development, where the models can be compiled to run on a target platform and data passed from MATLAB to the target for assessing the real-time capabilities of an algorithm and for fast, on-the-fly tuning of variables and parameters. All the aforementioned features combine to make MATLAB a powerful and flexible programming environment, providing rapid prototyping of image or video processing systems and easing their analysis including numerical issues. While graphical modeling of a system design is quite useful for quick prototyping, the constituent algorithms ultimately have to be converted back to text-based source code for actual deployment. Thus, algorithms can be coded in ".m" files but the process does not end there since MATLAB scripts are not, in general, suitable for real-time deployment.

Since MATLAB provides a rich text-based programming environment, it has the advantage that algorithms can be rapidly coded and their functionality verified. However, the features that enable the rapid development of algorithms also hinder their real-time implementation. The "interpretation of each instruction on-the-fly" characteristic of MATLAB in fact leads to a slow execution of ".m" source files, especially in loops where even though an instruction has already been interpreted in an iteration, it must be interpreted for every iteration. Since many image/video processing algorithms involve multiple nested loops, the overhead for interpretation of the source makes MATLAB unsuitable for real-time implementation. Although eliminating loops by modifying a MATLAB source code to take advantage of its built-in vector processing capabilities would help one to speed up simulations, such modifications might pose difficulties when directly porting the source code over to other general-purpose languages such as C/C++. Other strategies for speeding up MATLAB-based simulations using ".m" files would be to either compile the source code using the MATLAB compiler or to code time critical portions in C and link them to MATLAB using its MEX functionalities.

When interfacing MATLAB with C, it is important to note that MATLAB accesses image data in column-major ordering as opposed to row-major ordering commonly used in stand-alone C-based image processing routines. Hence, any C code developed to interface with MATLAB must carefully follow this data access convention. This inconsistency hinders the use of MATLAB interfaced with C.

It should be noted that such modifications are only meant to speed up simulations within the MATLAB environment and are not meant for real-time implementation. Also, since MATLAB is not strongly typed, transitioning from a MATLAB source to a general-purpose language for real-time deployment requires careful consideration of data-types. Using the Fixed-Point Toolbox can help ease the transition from a floating-point design to a fixed-point design, but despite these features, the path from MATLAB ".m" source files to standard general-purpose languages such as C/C++ is still quite involved considering that efficient MATLAB vectorized code usually has to be unvectorized or C-callable signal/image processing libraries have to be employed. It should be noted that tools have been developed to ease the transition from MATLAB to C, which could essentially automate the translation of MATLAB ".m" source files to C source files [157].

Unlike MATLAB, which was originally a text-only programming environment until Simulink was developed, LabVIEW was designed as a graphical-oriented programming environment from the start. LabVIEW provides powerful block-based system development. Combined with toolkits such as the IMAQ Vision and Advanced Signal Processing, LabVIEW can be used for rapid prototyping of a complete image/video processing system. The power of LabVIEW comes from its graphical-based programming environment that allows hierarchical and modular system design through so-called virtual instruments (VIs) and sub-VIs. LabVIEW can also be interfaced to use image processing algorithms written in general-purpose high-level languages such as C/C++ using the standard row-major array access convention. This feature helps ease the transition to a text-based source code for real-time deployment. The advantage of LabVIEW lies in the ease with which a graphic-user-interface can be used to adjust parameters in a simulation of an image or video processing system and to visualize intermediate and final results. LabVIEW can also be used for real-time assessment of an algorithm through hardware-in-the-loop type of development. Thus, similar to Simulink, LabVIEW can also be used to gain an understanding of the algorithms forming a complete system, but it is not generally meant for real-time use in a stand-alone product.

It should be noted that in some cases, C/C++ can be used for prototyping image and video processing algorithms, although it is usually used with MATLAB or LabVIEW or an image processing library for easy development. In such cases, usually the data-types used for variable declarations are not appropriate for real-time use. In many situations, floating-point data-types must be changed to appropriate integer data-types. While coding only in C/C++ from the start is beneficial from the standpoint of not having to translate the source code to a general-purpose, high-level language as in the case of MATLAB or LabVIEW, this also hinders data visualization in that data almost always have to be exported and visualized by another program. Another problem with coding in C/C++ from the start is foregoing the benefits of programming environments, such as MATLAB and LabVIEW, which can help to rapidly develop an algorithm.

Therefore, in practice, it is appropriate to utilize whatever combination of MATLAB, LabVIEW, and C/C++ for rapid development of an image or video processing system. It must be stressed that while MATLAB and LabVIEW are great tools for research and development, they are not meant to be used for real-time deployment in a stand-alone product. As such, it is necessary to translate a source code developed in MATLAB or LabVIEW into a standard, general-purpose, high-level programming language.

4.2.1.2 Real-Time Programming Languages

Depending on the underlying hardware platform, three types of programming languages can be employed for real-time image/video processing algorithms. They include general-purpose, high-level programming languages, hardware-description languages, and low-level assembly languages.

By far, the most widely used language for implementing real-time image/video processing algorithms on existing processors is the high-level C/C++ programming language. C is popular simply due to the fact that almost every processor has a compiler that can compile C source code into native machine code. That is to say, C is portable across a wide variety of machines. Due to its portability, many signal and image processing libraries exist that are optimized for different target platforms, easing the development of image/video processing algorithms. One general-purpose library, not optimized for any particular system, is the library provided in the <u>Numerical Recipes in C</u> [123], a standard reference that can be useful in porting MATLAB codes to C, especially for matrix-based computations [111]. C also has an advantage in its bit-level manipulation operations, and its ability to handle standard arrays, or arrays of objects in the case of C++, all of which are beneficial to developing efficient image/video processing algorithms [1]. That is why after the prototyping stage in MATLAB and/or LabVIEW, it is usually recommended to code an algorithm in C as a reference or baseline algorithm, which is then to be ported and specialized to a specific target platform.

Of course, if the hardware platform of interest is an FPGA, many hardware description languages (HDLs) exist for programming FPGAs, the most popular being VHDL. However, VHDL programming is often arcane to image/video processing algorithm developers, requiring a fundamentally different style of programming. New tools have been developed through which one can successfully go from a system designed in MATLAB straight to an FPGA implementation [156]. Such tools have the potential to extend the benefits of an FPGA-based solution within a faster development cycle.

While high-level software optimization techniques can be applied to the time critical or bottleneck portions of a C/C++ code to extract extra performance, under the assumption that such optimizations yield unsatisfactory results, the low-level assembly language of the target processor is required to be used in order to obtain the maximum performance. Proper use of assembly language requires an in-depth knowledge of the architecture of the processor, and is often time consuming to perform. Consequently, it is recommended to use assembly language if it is absolutely necessary and only on those portions of the code that are the most time critical.

4.2.2   Software Architecture Design

While translating a source code from a research development environment to a real-time environment is an involved task, it would be beneficial if the entire software system is well thought out ahead of time. Considering that real-time image/video processing systems usually consist of thousands of lines of code, proper design principles should be practiced from the start in order to ensure maintainability, extensibility, and flexibility in response to changes in the hardware or the algorithm [127]. Without a proper underlying structure, the entire system ends up becoming an unmanageable collection of source codes. Thus, it is critical to utilize good software engineering practices when developing a software-based real-time image/video processing system on programmable processors.

One key method of dealing with this problem is to make the software design modular from the start, which involves abstracting out algorithmic details and creating standard interfaces or application programming interfaces (APIs) to provide easy switching among different specific implementations of an algorithm [96]. Also beneficial is to create a hierarchical, layered architecture where standard interfaces exist between the upper layers and the hardware layer to allow ease in switching out different types of hardware so that if a hardware component is changed, only minor modifications to the upper layers will be needed.

Recently, there has been interest in applying the principles of object-oriented design patterns to aid in the development of real-time image/video processing systems. These methods help to promote software reuse and improve the ease with which new functionalities can be added to a system with minimal effort. Noting that research on such methods is still being performed, a proper design is expected to create a more efficient, compact, and easy to understand software architecture while not adversely affecting the performance of the system.

4.2.3   Real-time Operating System

In a real-time image/video processing system, certain tasks or procedures have strict real-time deadlines, while other tasks have firm or soft real-time deadlines. In order to be able to manage the deadlines and ensure a smoothly running system, it is useful to utilize a real-time operating system. Real-time operating systems allow the assignment of different levels of priorities to different tasks. With such an assignment capability, it becomes possible to assign higher priorities to hard real-time deadline tasks and lower priorities to other firm or soft real-time tasks. For portable embedded devices such as digital cameras, a real-time operating system can be used to free the upper layer application from managing the timing and scheduling of tasks, and handling file input/output operations [77]. Therefore, a real-time operating system is an important key component of the software of any practical real-time image/video processing system since it can be used to guarantee meeting real-time deadlines and thus ensuring deterministic behavior to a certain extent.

4.3   MEMORY MANAGEMENT

It is widely recognized by hardware designers that yearly increases in memory performance slowly lags behind such increases in computing performance. Since this trend shows no signs of stopping in the near future, it becomes important to carefully consider the management of memory resources in a real-time image/video processing system, especially when a vast amount of data must be dealt with. Due to the overwhelming importance of proper memory management, this section covers key concepts such as the basics of computer memory architecture, how image data is stored in memory, and several memory management optimization strategies.

4.3.1   Memory Performance Gap

Due to the ever-increasing gap between computation performance and memory performance, memory optimizations are becoming more critical than computation optimizations, considering that most algorithms when first ported are more memory-limited than compute-limited [26]. Important items of interest are seeking out methods to reduce cache-miss rates, to fetch only that portion of the entire image data that needs to be processed into the on-chip memory, to partition data, etc. Of course, another tool used for efficient movement of data across a system without CPU intervention is the direct memory access (DMA) feature found in modern processor architectures.

4.3.2   Memory Hierarchy

Most modern hardware architectures are outfitted with a hierarchical memory where each level is separated from the processor by increasing levels of access. As the access level to the processor increases, the memory size increases while the memory access speed decreases. The structure of a memory hierarchy is designed in such a way that to provide maximum memory access performance or throughput for frequently used program or data sections. Usually the hierarchy involves two cache levels, known as L1 cache and L2 cache, followed by external memories of varying sizes and speeds. The L1 cache is placed closest to the CPU core and is usually smaller than the L2 cache which is the next closest memory level to the CPU. Some architectures place these two levels directly on the processor, while in other architectures caches are physically separated from the processor. For instance, in some digital signal processors (DSPs), e.g., the TMS320C64x, there are separate L1 program and data caches and an L2 cache. The memory available on-chip normally provides the fastest access among various types of memories. As far as external memory is concerned, there are mainly two types that are used for real-time embedded systems: SRAMs and SDRAMs, where SRAMs provide lower memory access latencies and are thus "faster" than SDRAMs.

In image/video processing applications, it is beneficial to place the image being operated on within the on-chip memory to enable the processor to quickly access the necessary data with minimum latencies, reducing the overhead of external memory accesses. Since it is often the case that an entire image or video frame cannot fit within the available on-chip memory, the processing has to be reorganized or restructured to enable an efficient implementation on the target hardware. These issues are covered in Subsection 4.3.5 on memory optimization strategies.

4.3.3   Organization of Image Data in Memory

Considering that the handling of image data is one of the main difficulties for real-time implementations, it is important to understand how image data is represented in C/C++ and what are some commonly used methods for accessing them. In C/C++, image data is stored in row-major format that simply means that it is stored as an array of rows; this is different from MATLAB that stores image data in column-major format as an array of columns. Also, C/C++ uses zero-based indexing and square brackets to access array data, as opposed to MATLAB that uses 1s-based indexing and parenthetical brackets to access array data. Due to performance issues, images are usually not declared or accessed as two-dimensional (2D) arrays, but are declared and accessed using pointers that point to an address in memory where image data is stored in row-major format. Since a pointer points to image data in memory, a proper pointer dereferencing must be performed to access an image. Hence, the pixels of an $M \times N$ image $Img2d$ should be accessed as $Img2D[N*i + j]$ instead of $Img2D[i][j]$. This principle can be extended to higher dimensional data as well. For instance, elements of an $M \times N \times L$ three-dimensional (3D) image $Img3D$ can be accessed as $Img3D[N*L*i + L*j + k]$ instead of $Img3D[i][j][k]$.

4.3.4   Spatial Locality and Cache Hits/Misses

One restriction of the row-major storage format is the lack of spatial locality, meaning that pixels that are spatially local to each other in 2D are not stored very close to each other within memory [19, 26]. For instance, the horizontally adjacent pixels $Img2D[i][j]$ and $Img2D[i][j + 1]$ are separated only by a few bytes in memory, whereas the vertically adjacent pixels $Img2D[i][j]$ and $Img2D[i + 1][j]$ are separated by several bytes of pixel data in memory [19]. Consequently, the row-major storage of image data favors horizontal or row-wise memory accesses over vertical or columnwise memory accesses. This way there is usually no performance degradation for accessing image data along rows. On the other hand, accessing image data along columns poses serious performance degradations, creating ample opportunities for frequent cache misses, especially if image data reside in external memory.

A cache miss is defined as an attempted memory access by the processor, where the desired data is not located in the cache, forcing the processor to obtain the desired data from the slower, external memory. As mentioned earlier, image data being processed should be placed in the cache for increased performance not to cause cache misses noting that cache misses pose barriers to real-time implementations. Due to the fact that many low-level and intermediate level operations require access to neighboring pixels, this can be a grave source of performance loss. Mechanisms should be put into place in order to increase the number of cache hits over cache misses, where a cache hit is defined as an attempted memory access by the processor with the consideration that the desired data is already located within the cache.

4.3.5   Memory Optimization Strategies

While memory management optimization strategies could be regarded as software optimization strategies, a distinction is made here between the two because of the overwhelming importance of memory performance bottlenecks as opposed to computation bottlenecks. Memory optimizations are meant to alleviate memory performance bottlenecks, while software optimizations are meant to alleviate computation bottlenecks.

4.3.5.1   Internal Memory Versus External Memory

As mentioned previously, due to the faster access times afforded by on-chip memories, it is best to place frequently used items within internal memory to overcome the overhead of external memory accesses. Although it is desired to place an entire image into internal memory, it is often the case that the entire image does not fit into the on-chip memory. In such cases, it would be detrimental to just leave the image in external memory. Another important strategy to deal with this issue involves allocating a buffer section in the available internal memory, partitioning the image data into blocks the size of the allocated buffer, and performing processing on the smaller data blocks. Some important image data partitioning schemes include row-stripe partitioning and block partitioning. The most commonly used partitioning scheme is the row-stripe partitioning scheme where a few lines or rows of image data are prefetched to a buffer within the on-chip memory to enable faster data accesses. The fetching of a few lines to internal memory before any processing commences also has the benefit of reducing cache misses for operations, which require 2D spatial locality, since vertically adjacent pixels would now be located in the cache. Another partitioning scheme is to divide an image into either overlapping or nonoverlapping blocks depending on the type of processing being performed.

In addition to placing image data in internal memory, other frequently used items should also be placed in internal memory [70]. Since many embedded processors have internal program and data on-chip memories, critical portions of the code and other frequently used data items such as tables should also be considered for inclusion into on-chip memory as space permits. The benefits of on-chip memory over that of external memory cannot be stressed enough as efficient use and handling of image data and program code portions within on-chip memory is often critical to achieving a real-time performance.

4.3.5.2 Efficient Movement of Data

While making efficient use of available internal memory for storing image data is important for obtaining real-time performance, using precious CPU resources to perform the movement of data, for instance using the memcpy function, is not recommended [113, 114]. A key peripheral available in most modern processor architectures is the DMA controller, which can manage the movement of data without CPU assistance, leaving it free to focus on time critical computations rather than becoming engaged in data management. A DMA controller can usually manage multiple DMA channels simultaneously so that multiple data transfers can occur at the same time.

With the availability of DMA, efficient multibuffering strategies have been developed that allow concurrent processing and movement of data. As the name implies, multibuffering strategies make use of multiple buffers usually placed within on-chip memory to allow performing concurrent processing and movement of data. Depending on the type of processing being performed, usually three buffers are employed including *buffer* 1 and *buffer* 2 operating in the so-called "ping-pong" manner and *buffer* 3 operating as an output buffer. The scheme usually takes the form where a DMA channel is used to store a block of data in *buffer* 1, while processing proceeds on data in *buffer* 2 and results are placed in *buffer* 3. After processing on *buffer* 2 has been completed, the results in *buffer* 3 are sent out to external memory through a DMA channel, while processing proceeds with data in *buffer* 1 and another DMA channel is used to bring in another block of image data into *buffer* 2. Many variations of this scheme have been used, and some of them are detailed further in the examples section. An important and often overlooked issue regarding memory accesses is the alignment of data. DMA transfers can benefit from proper data alignment by maximizing the data bus throughput [91, 99].

4.3.5.3 Increasing Memory Access via Spatial and Temporal Locality

Another method of reducing slow external memory accesses is to move from an image-based processing scheme to a pixel-based processing scheme when multiple operations have to be performed on image data and there are no data dependencies between the operations [16]. An image-based processing scheme involves applying one operation to all the pixels and then applying another operation to all the pixels, etc. This is quite similar to the way MATLAB performs processing on images. A pixel-based processing scheme on the other hand is one that applies all the operations to one pixel, and the same is repeated for all the pixels.

The problem with an image-based processing scheme is that it does not make an efficient use of the cache memory scheme, since the same pixel would have to be read many times to complete the entire processing. In the pixel-based processing scheme, the pixel is read only once and all the operations are performed while the pixel resides in the internal on-chip memory. Thus, not only does pixel-based processing improve spatial and temporal locality of memory accesses, but also increases the computational intensity of the implementation, a measurement commonly used to gauge if an implementation is memory limited or not [143]. Computational intensity is defined as the ratio of the number of instructions executed to the number of memory accesses. If many instructions are being executed per memory access, then the coded routine is said to have a high computational intensity, while on the other hand if a small number of instructions are executed per memory access, then the coded routine is said to have a low computational intensity. In other words, a low computational intensity means that the coded routine is memory inefficient. Therefore, since more operations are performed per memory access in a pixel-based processing scheme, the use of such a scheme is beneficial when it is applicable.

4.3.5.4 Other Memory Optimization Methods

There are many other strategies that could be employed to achieve an efficient use of memory, and most of these indeed depend on the application, the available resources, and the critical bottlenecks. Various steps, such as using global variables instead of local variables to reduce the size of the stack, allocating enough memory to the heap for dynamic allocation, and taking advantage of packed data transfers to maximize memory bandwidth, should be considered when transitioning to a real-time implementation.

4.4   SOFTWARE OPTIMIZATION

After having ported a code to standard C as a reference for implementation on the target hardware, and applying/exploring compiler-level optimizations, software profiling should be performed to see what portions of the code pose significant bottlenecks for meeting the real-time requirements. Once determining the bottlenecks, steps need to be taken to optimize those portions of the code in order to bring the execution time within an acceptable real-time range. Use of image/video processing libraries, fixed-point arithmetic, software pipelining, and subword parallelism are popular techniques that are mentioned here.

4.4.1   Profiling

There is a certain established method of performing software optimizations to improve the efficiency of a source code. One should not proceed blindly in applying software optimization techniques but should be guided by certain accepted practices. The very first stage in the software optimization process is to profile the code, that is to say, gather the expended processor clock cycles or actual execution times of every function and subfunction in the code. Most modern integrated development environments include the capability to profile codes. Profiling the code reveals the portions that are the most time consuming. The goal of software optimizations is to apply code transformations to those time critical portions of the code in order to extract maximum performance from the underlying hardware architecture. In image/video processing, such time critical portions mostly involve nested loop statements. After applying a single transformation, the code should first be verified for functional accuracy, and then profiled once again. If after profiling, the code still does not achieve a satisfactory performance, the process needs to be repeated again; otherwise, the software optimization phase should be stopped, as there would be no benefit in seeking a faster performance than what is actually necessary.

4.4.2   Compiler Optimization Levels

Modern compilers are usually equipped with automatic software optimizers, which gather information from the code and attempt transformations on the code to make it more efficient. Often there are several levels of optimization, each one targeting a different aspect of optimization, with the highest one offering the ability to produce the most efficient code than the other levels. After applying compiler-level optimizations, the code should be profiled to see if any increases in performance were gained. In some cases, applying compiler optimization levels would actually degrade performance for some functions. Thus, if it is possible, compiler optimizations need to be applied only to those functions that experience an increase in performance. If the performance after applying compiler optimization levels is satisfactory, then no further optimization will be necessary, bearing in mind that this is usually not the case. Normally, it is best to proceed with applying the strategies discussed next to those remaining time critical portions of the code. An important item to note is the utilization of the `volatile` keyword to make sure that the compiler does not remove the variables it thinks are useless when they are actually needed for proper functionality. An example of this is having a flag variable in a hardware, interrupt-triggered interrupt service routine [124].

4.4.3   Fixed-Point Versus Floating-Point Computations and Numerical Issues

When porting over a standard C reference code to the target architecture, usually floating-point-based computations are kept intact at the first stages of optimization to verify the functionality. More often than not, floating-point computations pose a major performance bottleneck in real-time implementations, especially on fixed-point processors that are often used in embedded devices. Performing floating-point calculations on fixed-point processors is regarded as a waste of resources since floating-point calculations have to be emulated through software to properly handle the exponent and mantissa parts of the numbers. Considering that heavy computations are usually performed in loops, this performance degradation is magnified by the number of iterations through loops. In addition to this, compiler optimizations usually cannot be performed on loops with floating-point calculations on a fixed-point processor since the compiler has to add a function call to a library emulating floating-point calculations. A compiler would not optimize a loop with a function call within the loop [113, 114]. Therefore, performing floating-point calculations on fixed-point processors is not a good idea.

Fixed-point arithmetic is the preferred format of arithmetic in real-time image/video processing systems, especially in embedded systems. The reason behind the choice of fixed-point over floating-point is that fixed-point calculations are usually faster to perform since they can be accomplished using integers without the overhead imposed by floating-point calculations that have to take into consideration the mantissa and the exponent of a floating-point number representation. Also, noting that most embedded devices utilize fixed-point processors, it is quite inefficient to emulate floating-point arithmetic on such processors as this causes a considerable reduction in performance.

The speedup gained when transitioning from floating-point-based computations to fixed-point-based computations can be dramatic. Therefore, a transition to fixed-point arithmetic is essential toward achieving the highest levels of performance. However, in practice this is easier said than done, because the developer has to decide what the appropriate fixed-point format for the calculations should be. One of the most commonly known methods of coding fixed-point representation of numbers is the mixed integer fractional Q format, $Qm.n$, denoting an $m + n + 1$ bit number, where $m + 1$ bits to the left of the radix point constitute the 2's complement signed integer portion and n bits to the right of the radix point constitute the fractional portion [16, 124]. This format has the flexibility of allowing differing levels of accuracy to mitigate the effects of quantization on the numerical calculations. As a result, it becomes important to determine the necessary level of accuracy for each calculation and thus to use an appropriate fixed-point representation.

Numerical issues such as these and others including quantization effects, rounding, truncation, overflows, proper scaling, and the order of operations need to be carefully considered when implementing fixed-point computations. Multiplying two fractional numbers yields another fraction, but adding/subtracting two fractional numbers might generate numbers outside the fractional range causing overflow, hence the need for proper scaling [81]. The order of operations might also be critical in reducing the propagation of errors throughout a particular computation, for instance, in the calculation of local variances [124]. Also, care must be taken to keep track of the decimal point during the calculations. These issues can be rapidly and fully explored within the development environments such as MATLAB with its Fixed-point Toolbox or with Simulink, allowing a rapid determination of the required accuracy level.

It is recommended to find a representation that requires the least amount of bits in order to conserve memory and to enable use of packed data processing [16]. For example, using a 16-bit fixed-point representation instead of a 32-bit representation such as the popular $Q0.15$ format, or a 16-bit pure fractional fixed-point representation, allows representing numbers in the range $(-1, 1)$ [81, 124]. When implemented on a target platform, fixed-point computations are accomplished by first scaling data by an appropriate power of 2 via left bit-shifting to gain sufficient accuracy, performing the computations, and then scaling back down by the same power of 2 via right bit-shifting. Scaling by powers other than 2 is not time efficient since such scaling would often require a computationally expensive division operation in order to scale the result back to the original level [113, 114].

Some other important numerical issues involve performing division operations on fixed-point processors. In [81, 124], several strategies are suggested including using the repeated subtraction method, using a lookup table if the input range is known *a priori*, or using the Newton–Raphson algorithm.

4.4.4   Optimized Software Libraries

One should not discount the use of any dedicated image and signal processing libraries supplied by the manufacturer of the processor being used. Such libraries have been carefully crafted to extract maximum performance from the processor architecture and thus their use should be considered as a means of improving performance to meet real-time requirements. Of course, one should consult the supplied documentation on the available functions before using them. Some noteworthy libraries include Intel Integrated Performance Primitives Library for the Pentium line of processors, Texas Instruments DSPlib signal processing library, and Imglib image processing library. When used properly, these libraries can help one to cut down on the development time by not having to optimize certain basic functions that are used in many applications. While such libraries can be used to obtain a performance gain in certain cases, in other cases, it might be advantageous to specialize the implementation to the specific computation needed instead of utilizing a general-purpose function from the library.

4.4.5   Precompute Information

Sometimes certain computations are repeated over and over when in reality such computations are in fact just constants used within other computations. In such cases, it is beneficial to precompute constants and to store these in memory as a lookup table, trading the computation time with the time to access the constants in memory. Again, for frequently used constants, such data should be placed in internal memory if space permits or else in fast external memory. If the constants are always accessed from memory in a certain order, it would be helpful to store data in memory in that particular order to ease memory accesses.

4.4.6   Subroutines Versus In-Line Code

To improve performance, it is recommended to use in-line codes instead of subroutines [1]. While subroutines are usually considered to be a good software engineering practice, they incur overhead when called since variables have to be pushed onto a stack in order to be popped back upon return. To avoid such overheads, the function calls can be replaced with in-line codes. While this will increase the code size, the benefits gained in performance might outweigh the increase in the code size. This is another classic example of the space-time trade-off in software design.

4.4.7   Branch Predication

It is well known that control-intensive codes in their original forms are not amenable to parallel processing. One strategy of getting around this limitation is to employ the technique of branch predication that essentially converts control-intensive codes to data-intensive codes [91, 111]. As a result, such codes can be parallelized leading to improved performance.

4.4.8   Loop Transformations

Loops are usually the most critical portions of codes. As such, there are many strategies to help one to extract the most performance out of time critical loops both in the high-level and in the low-level (assembly) languages. It is often a good practice to first apply loop transformations in the high-level language before applying assembly optimizations.

The first step toward applying loop transformations is to carefully examine a loop to see if there are any unnecessary calculations within the loop itself. If such calculations are found, they should be removed. Calculations such as common subexpressions should usually be evaluated only once instead of multiple times. Hence, such calculations serve as candidates for removal.

Another important aspect of a loop is to avoid function calls within the loop. All function calls within a loop should be replaced with in-line codes. As mentioned before, function calls within a loop usually disqualify the loop from being examined for optimization by the compiler optimizer. Also, any floating-point operations within the loop should be removed when using a fixed-point processor since the compiler inserts software emulation to support floating-point computations. Thus, it is essential to remove any function calls within the loop. It should be noted that this does not apply to intrinsic functions, and C-callable functions supplied by the manufacturer that map directly to the assembly-level instructions of the target processor.

One of the most effective and commonly used high-level loop transformations is loop unrolling, which can be used to increase the instruction level parallelism (ILP) of the loop. Loop unrolling allows performing multiple iterations in one pass, grouping more computations together for simultaneous access to more data and thus reducing loop overhead due to branching and helping the compiler to create a more efficient scheduling of the main loop body. Of course, unrolling a loop increases the code size, but it can also lead to more efficient code. This is another classic example of the space-time trade-off in software design.

Software pipelining is a code optimization technique found useful for reducing the execution time of critical loops in an image/video processing algorithm. Essentially, this technique utilizes the ILP that is buried within the loops. As previously stated, a simple high-level form of increasing ILP is to perform loop unrolling, relying on the compiler to automatically schedule a software pipeline. If after profiling, satisfactory results are not obtained, then the pipeline must be hand scheduled based on the assembly language of the target hardware. This is a truly tedious task that requires analyzing a data-dependency graph and properly allocating processor resources while making sure no conflicts or data dependencies exist in instructions scheduled to run in parallel. Despite the tediousness of the task, the payoff of hand-scheduled pipelined assembly is to obtain the highest performance. There are some automatic tools available that can help relieve the burden of hand-coded software pipelining. It is recommended to use such tools if the only way to obtain the necessary level of performance is through a hand-scheduled software pipeline. Another option to consider before undertaking hand-scheduled software pipelining is to utilize compiler intrinsics that can be used to direct the compiler optimizer to produce a more efficient scheduling of a loop.

4.4.9   Packed Data Processing

Most modern architectures incorporate some form of packed data processing functionality to enhance the performance of processing data that is smaller than the width of the data path. Recall that pixel data is usually represented anywhere between 8 and 14 bits, depending on the accuracy required by the application. Since most modern data paths are of 32 bits, it is possible to process four 8-bit data or two 16-bit data pixels simultaneously, leading to processing more image data in one clock cycle. Such operations can be used to speed up matrix–vector computations frequently encountered in image/video processing algorithms. In order to use this packed data processing feature, the code usually has to be reorganized to implement the packed data processing mode, similar to code vectorizing [91]. It is important to exploit this feature to the fullest in order to process the maximum amount of image data in one clock cycle. This feature can be combined with software pipelining for even greater gains in performance. It is important to make sure data are properly aligned so that memory loads and stores operate with maximum efficiency. In the cases where it is necessary to access unaligned data, special instructions can be utilized, if available, to reduce the overhead involved from loading two adjacent data words and extracting the desired data via shifting and bit-masking operations [99].

4.5   EXAMPLES OF SOFTWARE METHODS

In what follows, several examples are presented to illustrate the deployment of the above software optimization tools for achieving real-time throughputs.

4.5.1   Software Design

Proper software design is essential for developing and maintaining real-time image/video processing systems. Several representative examples covering the design of software architectures and the use of object-oriented design principles are mentioned next.

4.5.1.1 Software Architectures

For real-time image/video processing systems, software architectures are just as important as hardware architectures. Modular and layered approaches have been used extensively in designing software architectures for real-time image/video processing systems as such approaches result in software that is easy to maintain and extend its future functionality. Here, three examples are mentioned to illustrate the use of modular and layered design and real-time operating system concepts.

One example incorporating both modular and layered approaches appears in [77], where a software architecture for an embedded multifunction digital camera system was discussed. This approach to software architecture design involved extracting application-specific features and device-dependent controls from the functional operation modules via a well-defined API and device driver interface (DDI). The architecture consisted of three layers, namely an application layer, a functional layer, and a system layer as well as two interfaces, the API and DDI. The application layer included a graphic-user-interface (GUI), while the functional layer implemented the supported camera functions. The system layer provided hardware abstraction, allowing reuse of all upper layer codes upon changes to hardware components. The system was managed using a real-time operating system that allowed the separation of upper level functionality from timing schedule details and file-allocation processes. As a result, this modular and layered approach allowed reusability and extended the programmability and flexibility of the software platform.

Another example of a layered approach can be seen in [13], where the software architecture of a system used for the automatic exposure control of a charge-coupled device camera was discussed. A layered approach to software architecture with five hierarchical layers was designed to allow a better isolation of different layers for development and validation purposes. The heart of the design revolved around the real-time layer that was constructed using RT-Linux, the hard real-time extension to the popular general-purpose Linux operating system. The use of this real-time operating system in the software architecture provided the ability to prioritize the hard real-time task of updating the exposure.

Another important aspect of software architecture design for real-time image/video processing systems is in the proper scheduling of tasks with different levels of priorities by utilizing a real-time operating system. The concept of multithreaded implementation has been gaining popularity in this regard. For example, in [18], a multithreaded software architecture was discussed that allowed a medical image enhancement procedure to be implemented in real-time. In terms of thread organization, an initial thread was responsible for the GUI and administrative control of the entire application. A frame acquisition/communication thread was used to handle the frame-grabber hardware and was given a high priority. Two threads for diagnostic data analysis and real-time enhancement also ran separately. In addition, a thread for storing data to the hard disk ran with a lower priority. The administrator thread managed the interaction among the different threads and had the ability to adjust priorities or halt threads if necessary. The use of such a software architecture was beneficial to maintain a real-time response, and to be reactive to the inputs by the user.

4.5.1.2 Object-Oriented Design Patterns

As previously mentioned, there has been an increased interest in the application of object-oriented design patterns in real-time image/video processing systems. One research group has primarily been involved in promoting the use of object-oriented design patterns for such systems as discussed in the references [90, 104, 127, 128].

In [90], an object-oriented design for a class of Kalman filters was presented that helped to decrease recoding efforts in adapting the filter implementation to different applications. By using the object-oriented programming design patterns named the "Gang of Four," a software implementation of the Kalman filter was achieved that was amenable to future extensions without much recoding effort. By making use of these design patterns, it was easy to abstract the detail of the filter implementation away, allowing a truly flexible software implementation supporting different numbers of inputs, different noise models, different filter implementations, and pixel- or blockwise processing modes.

Not only can object-oriented design patterns be used to limit recoding efforts in deploying differing versions of the same algorithm, they can also be used to limit the recoding efforts for changes in hardware. In [104], it was discussed how object-oriented design patterns could be applied to a real-time image processing problem, enabling and promoting software architecture reuse in the face of both algorithm and hardware alterations. Since real-time image/video processing systems consist of many lines of code, it would be beneficial to not have to redevelop the software architecture from scratch every time for a new design. Object-oriented design patterns can be used to help cut down on the development time by allowing the reuse of already tested and developed software for similar image/video processing problems.

Originally, it was thought that using such design patterns might have a detrimental effect on the real-time performance of a system due to the use of several layers of abstraction involved to create software reusable systems. In [127], it was shown that this was not the case. A study was performed which showed which applying object-oriented design patterns could produce well-designed and efficient image processing systems with easy extensibility, maintainability, and reuse without sacrificing real-time performance. Due to these findings, in [128], it was argued that real-time image processing applications should be written using an object-oriented approach to achieve efficient, maintainable, and understandable code from the start. Such a design approach was shown to be superior since it can eliminate inefficiencies in software that are common in other design approaches.

4.5.2   Memory Management

4.5.2.1   Increasing Spatial/Temporal Locality

There have been several examples in the literature on methods to increase the spatial or temporal locality of memory accesses for gaining speedups in processing. Here, five representative examples are mentioned that illustrate how restructuring the processing can lead to efficient use of the memory hierarchy necessary for achieving real-time performance.

As mentioned previously, the measure of computational intensity can be used as an indication of memory bottlenecks in a given computation. An example of using this measure can be seen in [143], where the problem of using the Discrete Wavelet Transform (DWT) for lossless compression of medical images was considered. The DWT approach chosen was the standard method based on a quadrature mirror filterbank. It was found that the implemented convolution source code had a low computational intensity, indicating that the source code was memory inefficient. To overcome the memory bottleneck, the computations were restructured in such a way that to reduce the number of external memory accesses per convolution operation. This had the effect of raising the computational intensity of the convolution source code and thus helped to achieve more efficient use of the memory hierarchy.

In most cases, the limited size of fast internal memories forces the restructuring of the processing to make an efficient use of the memory hierarchy. For example, in [30], the problem of designing an efficient image processing library within the strict memory constraints of an embedded digital camera platform was considered. The concept of "band processing" was utilized that involved partitioning an image into several smaller-sized data bands, sequentially processing the bands using a pipeline of band-based operators, and then collecting the bands into a single-output band. Such an approach provided a means of efficiently using the limited memory resources of the embedded platform.

Another example in which the processing was restructured for an efficient use of memory resources is reported in [31], where a memory-efficient algorithm was devised for the DWT calculations based on the overlapped block-transferring scheme and reorganization of the calculations. It was shown that the chosen algorithm for computing the wavelet coefficients suffered from the cache-miss problem in the filtering operation. Despite the computational efficiency of the algorithm, the cache-miss problem caused a poor performance during vertical filtering. An overlapped block-transferring method was thus devised to overcome the cache-miss problem. The method involved partitioning the image residing in the external memory into blocks the size of the cache. The processing was then restructured to allow the elimination of the cache-misses during vertical filtering. To overcome the issue of needing adjacent block information during this filtering procedure, horizontally overlapped blocks were considered.

In contrast to the traditional block- and strip-partitioning schemes often used in the reorganization of computations, a new data organization called "super-line" processing was developed in [44, 58, 101] specifically addressing the issue of cache-misses in algorithms that use multiresolution representations of image data. This organization involved dividing an image into partitions called "super-lines," where each partition contained all the necessary information from each level of the multiresolution representation to allow the application of the algorithm in one pass. In [44], the results showed that after application of the super-line approach, only 0.2% of the memory accesses were to the external memory, which helped to achieve a processing rate of 43.6 fps for 512×512 images on a desktop GPP. Also, in [101], the results showed that the super-line approach was able to reduce the miss rate from 99% to 0.8%, achieving a processing rate of 44 fps for 555×382 images on a desktop GPP.

4.5.2.2 Efficient Movement of Data

Due to the vast amount of data needed to be processed in real-time image/video processing systems, efficient movement of data plays an important role in obtaining real-time performance. DMA has been used extensively to allow efficient movement of data, and several buffering schemes have been developed for this purpose. For example, in [99], a data flow management scheme using a programmable DMA controller was discussed. Since the size of the on-chip data cache was limited and not large enough to hold an entire image frame, the images were processed in smaller chunks at a time. An on-chip programmable DMA controller was used to manage the movement of data concurrently with the computations, preventing the processor stalls introduced by waiting for the data to be fetched from the external memory to the on-chip memory. A double buffering scheme was employed involving the use of four buffers, two for input blocks (*ping_in_buffer* and *pong_in_buffer*) and two for output blocks (*ping_out_buffer* and *pong_out_buffer*), which were allocated in the on-chip memory. While the core processor was accessing *pong_out_buffer*, the DMA controller moved the previously calculated output block in *ping_out_buffer* to the external memory and brought the next input block from the external memory into *ping_in_buffer*. When the computation and data movements were completed, the processor and DMA controller switched buffers, and the processor started to use the ping buffers, while the DMA used the pong buffers.

Several other examples on the use of DMA for real-time image/video processing applications are reported in [31, 46, 77, 98, 138].

4.5.3   Software Optimization

There are many examples of software optimization techniques encountered in the literature. Next, several representative examples on the use of fixed-point computations, software libraries, and loop transformations are given.

4.5.3.1 Fixed-Point Computations

As most embedded platforms used in real-time image/video processing applications include fixed-point processors, reformulating floating-point computations in fixed-point is an important task for achieving real-time performance. For instance, in [11], it was found that floating-point computations were the major bottleneck toward reaching a real-time performance. The optimization strategy employed was to convert only the most computationally intensive parts of the algorithm over to fixed-point to ease the coding effort while still achieving a significant speedup.

Many researchers have used the $Qm.n$ fixed-point format in transitioning their algorithms from development to implementation. For instance, in [16], the $Q9.6$ fixed-point format was employed in the algorithm utilized for a real-time video analysis traffic surveillance application. The range and precision required were first verified in MATLAB before the implementation on the embedded platform. In addition, all the arithmetic operations used were implemented using the chosen fixed-point format. Another example illustrating the use of such a format appeared in [99], where the mapping of two-dimensional convolution on a VLIW media processor was considered. It was pointed out that generalized convolution required the normalization of the outcome via an expensive division operation. To avoid the division, a scaling factor was introduced into the filter kernel coefficients. Each coefficient was thus represented in the $Q0.15$ fixed-point format.

To support varying levels of dynamic range throughout the course of a multistage computation, a fixed-point format with a varying integer part can be used. An example of such an approach is seen in [143], where in order to deal with the dynamic range changes in the DWT (an increase from scale $j - 1$ to j in the forward DWT and a decrease from scale j to $j - 1$ in the inverse DWT), a 32-bit word length was used with a variable integer part. In order to determine the required word length, the propagation of errors from each scale was analyzed to determine the lower and upper bound errors and thus the required word length.

4.5.3.2 Software Libraries

Software libraries are used extensively in real-time image/video processing systems. Such libraries can aid in porting algorithms from a research development environment onto a target hardware platform.

While general-purpose libraries can be used as a first step toward porting, libraries optimized for the target hardware platform can provide an easy means of extracting a high level of performance from the platform without having to have a deep understanding of the underlying hardware architecture. For example, in [132], an algorithm for adaptive image fusion was originally developed in MATLAB using its Image Processing Toolbox. Since MATLAB is an interpreted language, it provided a rapid prototyping, but it was not suitable for real-time deployment. Thus, the MATLAB algorithm was ported over to the standard general-purpose C++ programming language. To simplify the porting effort and to help maximize the use of processor resources, the optimized Intel Image Processing Library was used to replace the MATLAB Image Processing Toolbox functions. Another example of the use of optimized libraries is covered in [146], where the optimized assembly code libraries supplied by the chip manufacturer were used to reduce the coding effort and to speed up the performance.

4.5.3.3 Loop Transformations

The most time-consuming portions of image/video processing algorithms often consist of low-level or intermediate-level operations that are implemented as nested loops, requiring multiple iterations to complete the processing for an entire image frame. The main techniques for speeding up these critical core loops include loop unrolling and software pipelining.

For example, in [99], it was noted that certain instructions took multiple clock cycles to complete, which led to latencies. Loop unrolling and software pipelining were used to cope with such latencies. Loop unrolling was able to fill the delay slots due to instruction latencies by taking advantage of single-cycle instructions via unrolling those instructions that accessed multiple data within the loop. Software pipelining was then used to overlap the operations from different iterations of the loop, allowing a parallel execution of loading and computation. Software pipelining required a prolog part to set up the loop kernel and an epilog part for finishing the computations. Loop unrolling and software pipelining were combined to increase the data processing throughput of the algorithm for obtaining a real-time performance.

Because writing a software pipeline code in assembly language is a cumbersome task, tools have been created to help assist in this process. One example of such a tool is mentioned in [49] for VLIW DSPs. In this example, the software pipeline optimization tool (SPOT) was used which combined a graphical schedule editor with an automatic conflict analyzer. The schedule editor provided a clear 2D visualization of the scheduled software pipeline with rows representing the multiple functional units of the VLIW DSP and the columns representing processor clock cycles. The conflict analyzer provided an automatic allocation of the processor registers in addition to an instant feedback on any data dependencies and other coding errors. This tool was also capable of generating the assembly source code file of the conflict-free scheduled pipeline. The results showed that the tool was able to achieve a speedup by a factor of 20 over the optimized C code. Basically, it provided a fast, simplified, and cost-effective method for including optimized hand-scheduled pipeline code in the workflow of developing a real-time image processing system.

4.6   SUMMARY

This chapter has covered the major topics regarding software methods for achieving real-time performance including software design, memory management, and software optimization. It should be stressed that before performing any software level modifications, a thorough examination of the algorithm involved should be carried out to identify key implementation details such as the required dynamic range and input/output data accuracy, the memory requirements, and any potential parallelism inherent in the algorithm. It is also essential to have a working knowledge of the hardware in case the assembly language programming needs to be used to extract the maximum performance out of the hardware. It is worth pointing out that many of the software methods mentioned in this chapter are standard methods and have been used extensively to optimize algorithms running on various hardware platforms toward obtaining real-time performance. Also, since the application of these methods is quite straightforward, it can be assured that careful application of these methods will help one to acquire most performance out of a given hardware platform.

CHAPTER 5

The Road Map

The major aim of this book has been to provide general guidelines toward the development of real-time image/video processing systems. Such guidelines have not previously appeared in *one place* as discussed here. In Chapter 2, several key strategies for designing simplified algorithms for real-time use were presented. In Chapter 3, an overview of currently available hardware implementation platforms including digital signal processors (DSPs), field programmable gate arrays (FPGAs), general-purpose processors (GPPs), and graphics processing units (GPUs) was given. Finally, in Chapter 4, major software methods for obtaining real-time performance were covered. In all these chapters, representative examples from the recent literature were carefully selected and presented to provide relevant real-world problems to illustrate the key concepts one needs to be aware of when transitioning to a real-time implementation. After having covered the algorithmic, hardware, and software aspects, a recommended road map is mentioned in this final chapter.

5.1 RECOMMENDED ROAD MAP

The following steps are considered to constitute the road map when taking an image/video processing algorithm to a real-time environment running on a hardware platform:

- **Step 1:** *Understand the algorithm*
  It is imperative that a deep understanding of the algorithm is first obtained beyond the high-level knowledge that is often adequate in a research environment. In this initial step, any algorithmic simplifications such as those covered in Chapter 2 should be carefully considered. This understanding also determines what hardware platform is most suitable for its real-time deployment.

- **Step 2:** *Port the algorithm to a reference C/C++ implementation*
  After having verified the algorithm in the research environment, it is often necessary to create a reference C implementation, not specialized to any particular hardware. This step is needed for debugging the transition from MATLAB or LabVIEW to C and

for providing a platform-independent version of the algorithm, which could be ported over to different hardware platforms. The output of the ported algorithm should be verified using the initial version.

- **Step 3:** *Understand the hardware*
  In addition to having a deep understanding of the algorithm, it is important to have a deep understanding of the underlying hardware architecture in order to maximize the available computational resources of the architecture involved. It is also useful to become acquainted with any compiler intrinsic instructions as well as any optimized image processing libraries during this step.

- **Step 4:** *Port the reference algorithm to the target hardware*
  This step involves porting the reference algorithm over to the target hardware platform.

- **Step 5:** *Profile and identify the bottlenecks*
  After having ported the reference algorithm to the target hardware, next step is to profile the algorithm to identify where the bottlenecks lie. One should also make use of the integrated development environment (IDE) of the target hardware during this step.

- **Step 6:** *Apply memory and high-level software optimizations*
  Once the time critical portions of the code are identified, next step includes memory optimizations and high-level software optimizations. Each modification of the code should be followed by a verification procedure to make sure that the outcome is as expected.

- **Step 7:** *Apply low-level software optimizations if necessary*
  After exhausting all high-level software optimizations, if the performance is still lacking, then one should resort to low-level software optimizations such as writing a hand-scheduled software pipeline assembly.

- **Step 8:** *Testing*
  Having achieved the desired performance, rigorous testing should be performed on the system to guarantee a smooth operation under worst-case conditions. The system should be put through a stress test to reveal any weak points to be patched up.

5.2   EPILOG

In this chapter, a recommended road map for the journey from research to reality was given. Although this book has presented many guidelines to assist in this journey, ultimately it is up to the system designer to select an appropriate collection of the presented guidelines in a particular real-time image/video processing application of interest.

References

[1] J. Ackenhusen, *Real-Time Signal Processing: Design and Implementation of Signal Processing Systems.* Englewood Cliffs, NJ: Prentice-Hall, 1999.

[2] M. Akil, "Case Study of a Dynamically Reconfigurable Architecture Implementing a Watershed Segmentation," *Proceedings of SPIE-IS&T Electronic Imaging Conference on Real-Time Imaging*, SPIE Vol. 4666, pp. 133–140, March 2002.

[3] M. Akil, "Dedicated Architecture for Topological Operators for Grayscale Image Processing," *Proceedings of SPIE-IS&T Electronic Imaging Conference on Real-Time Imaging*, SPIE Vol. 5012, pp. 75–82, April 2003.

[4] M. Akil, "Architecture for Hardware Thinning and Crest Restoration in Graylevel Images," *Proceedings of SPIE-IS&T Electronic Imaging Conference on Real-Time Imaging*, SPIE Vol. 5297, pp. 242–253, May 2004.

[5] M. Akil, "ARM-Based Embedded Processor: Real-time Implementation for Thinning and Crest Restoration in Gray-Level Images," *Proceedings of SPIE-IS&T Electronic Imaging Conference on Real-Time Imaging*, SPIE Vol. 5671, pp. 102–110, February 2005.

[6] A. Amer, "Memory-Based Spatio-Temporal Real-Time Object Segmentation for Video Surveillance," *Proceedings of SPIE-IS&T Electronic Imaging Conference on Real-Time Imaging*, SPIE Vol. 5012, pp. 10–21, April 2003.

[7] M. Arias-Estrada and J. Xicotencatl, "Real-Time FPGA-Based Architecture for Stereo Vision," *Proceedings of SPIE-IS&T Electronic Imaging Conference on Real-Time Imaging*, SPIE Vol. 4303, pp. 59–66, April 2001.

[8] A. Atsalakis, N. Papamarkos, N. Kroupis, D. Soudris, and A. Thanailakis, "Colour Quantization Technique Based on Image Decomposition and its Embedded System Implementation," *Proceedings of IEE Vision, Image, and Signal Processing*, Vol. 151, No. 6, pp. 511–524, December 2004.doi:10.1049/ip-vis:20040552

[9] N. Baba, H. Matsuo, and T. Ejima, "HeadFinder: A Real-Time Robust Head Detection and Tracking System," *Proceedings of SPIE-IS&T Electronic Imaging Conference on Real-Time Imaging*, SPIE Vol. 4666, pp. 42–51, March 2002.

[10] J. Batlle, J. Marti, P. Ridao, and J. Amat, "A New FPGA/DSP-Based Parallel Architecture for Real-Time Image Processing," *Journal of Real-Time Imaging*, Vol. 8, No. 5 pp. 345–356, October 2002.doi:10.1006/rtim.2001.0273

[11] A. Batur, B. Flinchbaugh, and M. Hayes III, "A DSP-Based Approach for the Implementation of Face Recognition Algorithms," *Proceedings of the IEEE International Conference on Acoustics, Speech, and Signal Processing*, Vol. 2, pp. 253–256, April 2003.

[12] R. Blahut, *Fast Algorithms for Digital Signal Processing*. Reading, MA: Addison-Wesley, 1985.

[13] S. Borromeo and J. Aparicio, "Real-Time Implementation of a Control System for Exposure Time of CCD," *Proceedings of SPIE-IS&T Electronic Imaging Conference on Real-Time Imaging*, SPIE Vol. 5671, pp. 44–51, February 2005.

[14] E. Bourennane, C. Milan, M. Paindavoine, and S. Bouchoux, "Real-Time Image Rotation Using Dynamic Reconfiguration," *Journal of Real-Time Imaging*, Vol. 8, No. 4, pp. 277–289, April 2002.doi:10.1006/rtim.2002.0274

[15] A. Bovik, "Introduction to Digital Image and Video Processing," in *Handbook of Image & Video Processing*, A. C. Bovik, Ed. Amsterdam: Elsevier Academic Press, 2005.

[16] M. Bramberger, J. Brunner, B. Rinner, and H. Schwabach, "Real-Time Video Analysis on an Embedded Smart Camera for Traffic Surveillance," *Proceedings of the 10th IEEE Real-Time and Embedded Technology and Applications Symposium*, pp. 25–28, May 2004.

[17] M. Bramberger, *Distributed Dynamic Task Allocation in Clusters of Embedded Smart Cameras*, Ph.D. Dissertation, Graz University of Technology, Austria, 2005.

[18] J. Bredno, B. Martin-Leung, and K. Eck, "Software Architecture for Live Enhancement of Medical Images," *Proceedings of SPIE-IS&T Electronic Imaging Conference on Real-Time Imaging*, SPIE Vol. 5297, pp. 122–133, May 2004.

[19] K. Brifault and H. Charles, "Data Cache Management on EPIC Architecture: Optimizing Memory Access for Image Processing," *ACM SIGARCH Computer Architecture News*, Vol. 32, No. 3, pp. 35–42, June 2004.doi:10.1145/1024295.1024300

[20] H. Broers, W. Caarls, P. Jonker, and R. Kleihorst, "Architecture Study for Smart Cameras," *Proceedings of the European Optical Society Conference on Industrial Imaging and Machine Vision*, pp. 39–49, June 2005.

[21] C. Bruyns and B. Feldman, "Image Processing on the GPU: A Canonical Example," *Computer Architecture Course Project*, Department of Computer Science, University of California Berkeley, Fall 2003.

[22] A. Carter and N. Audsley, "Architecture for Dynamically Reconfigurable Real-Time Lossless Compression," *Proceedings of SPIE-IS&T Electronic Imaging Conference on Real-Time Imaging*, SPIE Vol. 5297, pp. 231–241, May 2004.

[23] C. Castro-Pareja, J. Jagadeesh, and R. Shekhar, "FPGA-Based Acceleration of Mutual Information Calculation for Real-Time 3D Image Registration," *Proceedings of SPIE-IS&T Electronic Imaging Conference on Real-Time Imaging*, SPIE Vol. 5297, pp. 212–219, May 2004.

[24] C. Castro-Pareja, O. Dandekar, and R. Shekhar, "FPGA-Based Real-Time Anisotropic Diffusion Filtering of 3D Ultrasound Images," *Proceedings of SPIE-IS&T Electronic Imaging Conference on Real-Time Imaging*, SPIE Vol. 5671, pp. 123–131, February 2005.

[25] E. Cabello, M. Sánchez, and J. Delgado, "A New Approach to Identify Big Rocks with Application to the Mining Industry," *Journal of Real-Time Imaging*, Vol. 8, No. 1, pp. 1–9, February 2002.doi:10.1006/rtim.2000.0255

[26] H. Chen, K. Li, and B. Wei, "Memory Performance Optimizations for Real-Time Software HDTV Decoding," *Journal of VLSI Signal Processing*, Vol. 41, No. 2, pp. 193–207, September 2005.doi:10.1007/s11265-005-6650-7

[27] L. Chen and J. Li, "Content-Based Image Classification Using Quasi-Gabor Filters," *Proceedings of SPIE-IS&T Electronic Imaging Conference on Real-Time Imaging*, SPIE Vol. 5297, pp. 287–293, May 2004.

[28] L. Chen and C. Grecos, "Fast Skin Color Detector for Face Extraction," *Proceedings of SPIE-IS&T Electronic Imaging Conference on Real-Time Imaging*, SPIE Vol. 5671, pp. 93–101, February 2005.

[29] T. Chen and K. Chung, "A New Randomized Algorithm for Detecting Lines," *Journal of Real-Time Imaging*, Vol. 7, No. 6, pp. 473–481, December 2001.doi:10.1006/rtim.2001.0233

[30] A. Chihoub, Y. Bai, and V. Ramesh, "A Band Processing Imaging Library for a TriCore-Based Digital Still Camera," *Journal of Real-Time Imaging*, Vol. 7, No. 4, pp. 327–337, August 2001.doi:10.1006/rtim.2000.0244

[31] B. Choi, K. Choi, M. Hwang, J. Cho, and S. Ko, "Real-Time DSP Implementation of Motion-JPEG2000 Using Overlapped Block Transferring and Parallel-Pass Methods," *Journal of Real-Time Imaging*, Vol. 10, No. 5, pp. 277–284, October 2004.

[32] G. Conte, S. Tommesani, and F. Zanichelli, "The Long and Winding Road to High-Performance Image Processing with MMX/SSE," *Proceedings of the 5th IEEE International Workshop on Computer Architectures for Machine Perception*, pp. 302–310, September 2000.

[33]   P. Dang, T. Nguyen, and T. Tran, "High-Performance Low-Power BinDCT Coprocessor for Wireless Video Applications," *Proceedings of SPIE-IS&T Electronic Imaging Conference on Real-Time Imaging*, SPIE Vol. 5297, pp. 254–263, May 2004.

[34]   E. Davies, "A Sampling Approach to Ultra-Fast Object Location," *Journal of Real-Time Imaging*, Vol. 7, No. 4, pp. 339–335, August 2001.

[35]   E. Davies, *Machine Vision: Theory, Algorithms, Practicalities*. San Francisco, CA: Morgan Kauffmann Publishers, 2005.

[36]   X. Desurmont, J. Delaigle, A. Bastide, and B. Macq, "A Generic Flexible and Robust Approach for Intelligent Real-Time Video-Surveillance Systems," *Proceedings of SPIE-IS&T Electronic Imaging Conference on Real-Time Imaging*, SPIE Vol. 5297, pp. 134–141, May 2004.

[37]   X. Desurmont, B. Lienard, J. Meessen, and J. Delaigle, "Real-Time Optimizations for Integrated Smart Network Camera," *Proceedings of SPIE-IS&T Electronic Imaging Conference on Real-Time Imaging*, SPIE Vol. 5671, pp. 85–92, February 2005.

[38]   S. Dhanani, "FPGAs Enabling Consumer Electronics – A Growing Trend," *FPGA and Programmable Logic Journal*, http://www.fpgajournal.com/, 2005.

[39]   O. Djekoune and K. Achour, "Incremental Hough Transform: An Improved Algorithm for Digital Device Implementation," *Journal of Real-Time Imaging*, Vol. 10, No. 6, pp. 351–363, December 2004.

[40]   S. Dockstader and A. Tekalp, "On the Tracking of Articulated and Occluded Video Object Motion," *Journal of Real-Time Imaging*, Vol. 7, No. 5, pp. 415–432, October 2001.

[41]   K. Dong, M. Hu, Z. Ji, and B. Fang, "Research on Architectures for High Performance Image Processing," *Proceedings of the Fourth International Workshop on Advanced Parallel Processing Technologies*, September 2001.

[42]   E. Dougherty and P. Laplante, *Introduction to Real-time Imaging*. Bellingham, WA/Piscataway, NJ: SPIE Press/IEEE Press, 1995.

[43]   A. Downton and D. Crookes, "Parallel Architectures for Image Processing," *Electronics & Communication Engineering Journal*, Vol. 10, No. 3, pp. 139–151, June 1998.

[44]   K. Eck, H. Fillbrandt, G. Kiefer, and T. Aach, "Implementation of a Nonlinear Gradient Adaptive Filter for Processing of Large-Size Medical Sequences on General-Purpose Hardware," *Proceedings of SPIE-IS&T Electronic Imaging Conference on Real-Time Imaging*, SPIE Vol. 5012, pp. 61–70, April 2003.

[45] L. Estevez, "Real-Time Video Surveillance System Architecture," *Proceedings of SPIE-IS&T Electronic Imaging Conference on Real-Time Imaging*, SPIE Vol. 4303, pp. 19–26, April 2001.

[46] F. Faita, V. Gemignani, and M. Giannoni, "A Fully Customizable DSP Based System for Real-Time Imaging," *Proceedings of the Global Signal Processing Conference and Expo*, 2003.

[47] M. Feil and A. Uhl, "Real-Time Image Analysis Using MIMD Parallel à trous Wavelet Algorithms," *Journal of Real-Time Imaging*, Vol. 7, No. 6, pp. 483–493, December 2001.

[48] M. Fu, G. Jullien, V. Dimitrov, M. Ahmadia, and W. Miller, "The Application of 2D Algebraic Integer Encoding to a DCT IP Core," *Proceedings of the 3rd IEEE International Workshop on System-on-Chip for Real-Time Applications*, pp. 66–69, June/July 2003.

[49] J. Fuertler, K. Mayer, W. Krattenthaler, and I. Bajla, "Novel Development Tool for Software Pipeline Optimization for VLIW-DSPs Used in Real-Time Image Processing," *Proceedings of SPIE-IS&T Electronic Imaging Conference on Real-Time Imaging*, SPIE Vol. 5012, pp. 132–143, April 2003.

[50] J. Fung, "Computer Vision on the GPU," in *GPU Gems 2: Programming Techniques for High-Performance Graphics and General-Purpose Computation*, Matt Pharr, Ed. Reading, MA: Addison-Wesley, 2005, pp. 649–666.

[51] F. Gallegos-Funes and V. Ponomaryov, "Real-Time Image Filtering Scheme Based on Robust Estimators in Presence of Impulsive Noise," *Journal of Real-Time Imaging*, Vol. 10, No. 2, pp. 69–80, April 2004.

[52] M. Gamadia, V. Peddigari, N. Kehtarnavaz, G. Cook, and S. Lee, "Real-Time Implementation of Autofocus on the TI DSC Processor," *Proceedings of SPIE-IS&T Electronic Imaging Conference on Real-Time Imaging*, SPIE Vol. 5297, pp. 10–18, May 2004.

[53] R. Garcia, J. Batlle, and J. Salvi, "A New Approach to Pose Detection Using a Trinocular Stereovision System," *Journal of Real-Time Imaging*, Vol. 8, No. 2, pp. 73–93, April 2002.

[54] A. Gentile, S. Vitabile, L. Verdoscia, and F. Sorbello, "Image Processing Chain for Digital Still Cameras Based on the SIMPil Architecture," *Proceedings of the International Conference on Parallel Processing*, pp. 215–222, June 1005.

[55] P. Golding, *Next Generation Wireless Application*. New Jersey: John Wiley & Sons, 2004.

[56] R. Gonzales and R. Woods, *Digital Image Processing*. Englewood Cliffs, NJ: Prentice-Hall, 2002.

[57] C. Grecos, A. Saparon, and S. Jones, "Novel Scanning Order for Fast Elimination of Candidate Predictors in MPEG-2 Block-Based Motion Estimation," *Proceedings*

of SPIE-IS&T Electronic Imaging Conference on Real-Time Imaging, SPIE Vol. 5012, pp. 71–74, April 2003.

[58] A. Groth and K. Eck, "Real-Time Implementation of a Multiresolution Motion-Compensating Temporal Filter on General-Purpose Hardware," *Proceedings of SPIE-IS&T Electronic Imaging Conference on Real-Time Imaging*, SPIE Vol. 5671, pp. 132–141, February 2005.

[59] N. Gupta and P. Sinha, "FPGA Implementation of Fuzzy Morphological Filters," *Proceedings of SPIE-IS&T Electronic Imaging Conference on Real-Time Imaging*, SPIE Vol. 5297, pp. 220–230, May 2004.

[60] N. Guy, *Photonotes Dictionary of Photography*, http://www.photonotes.org/, 2004.

[61] Y. Han, P. Song, H. Chung, and H. Hahn, "An Active Contour-Based SSD Algorithm for Tracking a Moving Object," *Proceedings of SPIE-IS&T Electronic Imaging Conference on Real-Time Imaging*, SPIE Vol. 5297, pp. 26–34, May 2004.

[62] C. Hentschel, R. Braspenning, and M. Gabrani, "Scalable Algorithms for Media Processing," *Proceedings of the International Conference on Image Processing*, Vol. 3, pp. 342–345, October 2001.

[63] J. Hoshino and M. Kourogi, "Fast Panoramic Image Mosaicing Using One-Dimensional Flow Estimation," *Journal of Real-Time Imaging*, Vol. 8, No. 2, pp. 95–103, April 2002.doi:10.1006/rtim.2001.0257

[64] M. Houston, *General Purpose Computation on Graphics Processors (GPGPU)*. Stanford University Graphics Lab, http://graphics.stanford.edu/~mhouston/, 2005.

[65] H. Hunter and J. Moreno, "A New Look at Exploiting Data Parallelism in Embedded Systems," *Proceedings of the International Conference on Compilers, Architectures, and Synthesis for Embedded Systems*, pp. 159–169, October/November 2003.

[66] S. Hussmann and T. Ho, "A High-Speed Subpixel Edge Detector Implementation Inside a FPGA," *Journal of Real-Time Imaging*, Vol. 9, No. 5, pp. 361–368, October 2003.doi:10.1016/j.rti.2003.09.013

[67] S. Hussmann and P. Deng, "High Speed Optical Mark Reader Hardware Implementation at Low Cost Using Programmable Logic," *Journal of Real-Time Imaging*, Vol. 11, No. 1, pp. 19–30, February 2005.

[68] G. Iannizzotto and L. Vita, "On-Line Object Tracking for Colour Video Analysis," *Journal of Real-Time Imaging*, Vol. 8, No. 2, pp. 145–155, April 2002.doi:10.1006/rtim.2001.0267

[69] A. Iketani, A. Nagai, Y. Kuno, and Y. Shirai, "Real-Time Surveillance System Detecting Persons in Complex Scenes," *Journal of Real-Time Imaging*, Vol. 7, No. 5, pp. 433–446, October 2001.doi:10.1006/rtim.2000.0211

[70] P. Jara, I. Garcia, and B. Usevitch, "Analysis and Optimization of JPEG2000 in the TMS320C6701," *Proceedings of the Global Signal Processing Conference and Expo*, 2003.

[71] A. Jerraya, H. Tenhunen, and W. Wolf, "Guest Editors' Introduction: Multiprocessor Systems-on-Chips," *IEEE Computer Magazine*, Vol. 38, No. 7, pp. 36–40, July 2005.

[72] Q. Ji and X. Yang, "Real-Time Eye, Gaze, and Face Pose Tracking for Monitoring Driver Vigilance," *Journal of Real-Time Imaging*, Vol. 8, No. 5, pp. 357–377, October 2002.doi:10.1006/rtim.2002.0279

[73] J. Jiang, J. Xia, and C. Hou, "Empirical Study of Partial Decoding for Fast Browsing of MPEG-2 Compressed Videos," *Proceedings of SPIE-IS&T Electronic Imaging Conference on Real-Time Imaging*, SPIE Vol. 4666, pp. 1–9, March 2002.

[74] J. Jin, Z. Zhu, and G. Xu, "Digital Video Sequence Stabilization Based on 2.5D Motion Estimation and Inertial Motion Filtering," *Journal of Real-Time Imaging*, Vol. 7, No. 4, pp. 357–365, August 2001.doi:10.1006/rtim.2000.0243

[75] S. Kang, J. Min, and J. Paik, "Multiple-Camera Tracking System with Seamless Object Handover," *Proceedings of SPIE-IS&T Electronic Imaging Conference on Real-Time Imaging*, SPIE Vol. 4303, pp. 27–34, April 2001.

[76] J. Kang, H. Ki, J. Jung, J. Shin, and J. Paik, "Hierarchical Active Shape Model for Real-Time Tracking of Nonrigid Objects," *Proceedings of SPIE-IS&T Electronic Imaging Conference on Real-Time Imaging*, SPIE Vol. 5297, pp. 55–65, May 2004.

[77] W. Kao, T. Sun, and S. Lin, "A Robust Embedded Software Platform for Versatile Camera Systems," *Proceedings of the IEEE International Symposium on Circuits and Systems*, Vol. 5, pp. 5015–5018, May 2005.

[78] N. Kehtarnavaz, H. Oh, and Y. Yoo, "Development and Real-Time Implementation of Auto White Balancing Scoring Algorithm," *Journal of Real-Time Imaging*, Vol. 8, No. 5, pp. 379–386, October 2002.doi:10.1006/rtim.2001.0287

[79] N. Kehtarnavaz and H. Oh, "Development and Real-Time Implementation of a Rule-Based Auto-Focus Algorithm," *Journal of Real-Time Imaging*, Vol. 9, No. 3, pp. 197–203, June 2003.doi:10.1016/S1077-2014(03)00037-8

[80] N. Kehtarnavaz, H. Oh, and Y. Yoo, "Color Filter Array Interpolation Using Color Correlations and Directional Derivatives," *Journal of Electronic Imaging*, Vol. 12, No. 4, pp. 621–632, October 2003.doi:10.1117/1.1607966

[81] N. Kehtarnavaz, *Real-Time Digital Signal Processing Based on the TMS320C6000.* Amsterdam: Elsevier, 2004.

[82] F. Kelly and A. Kokaram, "Fast Image Interpolation for Motion Estimation Using Graphics Hardware," *Proceedings of SPIE-IS&T Electronic Imaging Conference on Real-Time Imaging*, SPIE Vol. 5297, pp. 184–194, May 2004.

[83] L. Kessal, N. Abel, and D. Demigny, "Real-Time Image Processing with Dynamically Reconfigurable Architecture," *Journal of Real-Time Imaging*, Vol. 9, No. 5, pp. 297–313, October 2003.doi:10.1016/j.rti.2003.07.001

[84] K. Kim, J. Park, R. Kim, R. Park, S. Lee, and I. Kim, "Real-Time Implementation of the Relative Position Estimation Algorithm Using the Aerial Image Sequence," *Journal of Real-Time Imaging*, Vol. 8, No. 1, pp. 11–21, February 2002.doi:10.1006/rtim.2001.0261

[85] N. Kim and N. Kehtarnavaz, "DWT-Based Scene-Adaptive Color Quantization," *Journal of Real-Time Imaging*, Vol. 11, No. 5–6, pp. 443–453, October–December, 2005.

[86] R. Kirsch, "SEAC and the Start of Image Processing at the National Bureau of Standards," *IEEE Annals of the History of Computing*, Vol. 20, No. 2, pp. 7–13, 1998.doi:10.1109/85.667290

[87] G. Knowles, "Real-Time Hardware Architectures for the Bi-Orthogonal Wavelet Transform," *Proceedings of SPIE-IS&T Electronic Imaging Conference on Real-Time Imaging*, SPIE Vol. 5012, pp. 102–109, April 2003.

[88] L. Kotoulas and I. Andreadis, "Efficient Hardware Architectures for Computation of Image Moments," *Journal of Real-Time Imaging*, Vol. 10, No. 6, pp. 371–378, December 2004.doi:10.1016/j.rti.2004.09.002

[89] S. Kyo, S. Okazaki, and T. Arai, "An Integrated Memory Array Processor Architecture for Embedded Image Recognition Systems," *Proceedings of the 32nd International Symposium on Computer Architecture*, pp. 134–145, June 2005.

[90] P. Laplante and C. Neill, "A Class of Kalman Filters for Real-Time Image Processing," *Proceedings of SPIE-IS&T Electronic Imaging Conference on Real-Time Imaging*, SPIE Vol. 5012, pp. 22–29, April 2003.

[91] S. Larin, "Introduction to AltiVec™ Technology: Ten Easy Ways to Vectorize Your Code," *Freescale Technology Forum*, 2005.

[92] J. Lee, J. Ko, and E. Kim, "A Real-Time Face Detection and Tracking for Surveillance System using Pan/Tilt Controlled Stereo Camera," *Proceedings of SPIE-IS&T Electronic Imaging Conference on Real-Time Imaging*, SPIE Vol. 5297, pp. 152–162, May 2004.

[93] K. Lee, K. Wong, S. Or, and Y. Fung, "3D Face Modeling from Perspective-Views and Contour-Based Generic-Model," *Journal of Real-Time Imaging*, Vol. 7, No. 2, pp. 173–182, April 2001.doi:10.1006/rtim.2000.0241

[94] B. Likhterov and N. Kopeika, "Mixed-Signal Architecture for Real-Time Two-Dimensional Live TV Image Restoration," *Journal of Real-Time Imaging*, Vol. 7, No. 2, pp. 183–194, April 2001.doi:10.1006/rtim.2000.0237

[95] A. Lins and K. Williston, *Processors for Consumer Video Applications.* Berkeley Design Technology, http://www.bdti.com/, 2004.

[96] V. Loganthan, *Flexible and Scalable Movie Architecture for DSP Based DSC/DM Systems*, TechOnLine Publication, http://www.technonline.com/, October 2005.

[97] A. Lugmayr, R. Creutzburg, S. Kalli, and A. Tsoumanis, "Face Customization in a Real-Time DigiTV Stream," *Proceedings of SPIE-IS&T Electronic Imaging Conference on Real-Time Imaging*, SPIE Vol. 4666, pp. 18–29, March 2002.

[98] S. Mahlknecht, R. Oberhammer, and G. Novak, "A Real-Time Image Recognition System for Tiny Autonomous Mobile Robots," *Proceedings of the 10th IEEE Real-Time and Embedded Technology and Applications Symposium*, pp. 324–330, May 2004.

[99] R. Managuli, G. York, D. Kim, Y. Kim, "Mapping of Two-Dimensional Convolution on Very Long Instruction Word Media Processors for Real-Time Performance," *Journal of Electronic Imaging*, Vol. 9, No. 3, pp. 327–335, July 2000.doi:10.1117/1.482755

[100] J. Martinez, E. Costa, P. Herreros, X. Sànchez, and R. Baldrich, "A Modular and Scalable Architecture for PC-Based Real-Time Vision Systems," *Journal of Real-Time Imaging*, Vol. 9, No. 2, pp. 99–112, April 2003.doi:10.1016/S1077-2014(03)00002-0

[101] M. Meribout, M. Nakanishi, and T. Ogura, "Accurate and Real-Time Image Processing on a New PC-Compatible Board," *Journal of Real-Time Imaging*, Vol. 8, No. 1, pp. 35–51, February 2002.doi:10.1006/rtim.2001.0269

[102] J. Molineros and R. Sharma, "Real-Time Tracking of Multiple Objects Using Fiducials for Augmented Reality," *Journal of Real-Time Imaging*, Vol. 7, No. 6, pp. 495–506, December 2001.doi:10.1006/rtim.2001.0242

[103] R. Neapolitan and K. Naimipour, *Foundations of Algorithms Using Java Pseudocode.* Boston, MA: Jones and Bartlett Publishers, 2004.

[104] C. Neill and P. Laplante, "Imaging Frameworks: Design for Reuse in Real-Time Imaging," *Proceedings of SPIE-IS&T Electronic Imaging Conference on Real-Time Imaging*, SPIE Vol. 5297, pp. 1–9, May 2004.

[105] M. Nguyen, "Distance-Invariant Object Recognition by Real-Time Vision," *Proceedings of SPIE-IS&T Electronic Imaging Conference on Real-Time Imaging*, SPIE Vol. 5012, pp. 30–36, April 2003.

[106] Y. Nishikawa, S. Kawahito, and T. Inoue, "Parallel Image Compression Circuit for High-Speed Camera, *Proceedings of SPIE-IS&T Electronic Imaging Conference on Real-Time Imaging*, SPIE Vol. 5671, pp. 111–122, February 2005.

[107] H. Oh and N. Kehtarnavaz, "DSP-Based Automatic Color Reduction Using Multiscale Clustering," *Proceedings of SPIE-IS&T Electronic Imaging Conference on Real-Time Imaging*, SPIE Vol. 4303, pp. 79–88, April 2001.

[108] H. Oh, N. Kehtarnavaz, Y. Yoo, S. Reis, and R. Talluri, "Real-Time Implementation of Auto White Balancing and Auto Exposure on the TI DSC Platform," *Proceedings of SPIE-IS&T Electronic Imaging Conference on Real-Time Imaging*, SPIE Vol. 4666, pp. 52–56, March 2002.

[109] J. Orwell, P. Remagnino, and G. Jones, "Optimal Color Quantization for Real-Time Object Recognition," *Journal of Real-Time Imaging*, Vol. 7, No. 5, pp. 401–414, October 2001.doi:10.1006/rtim.2000.0209

[110] J. Owens, D. Luebke, N. Govindaraju, M. Harris, J. Kruger, A. Lefohn, and T. Purcell, "A Survey of General-Purpose Computation on Graphics Hardware," *Eurographics State of the Art Reports*, pp. 21–51, August 2005.

[111] I. Ozer, T. Lu, and W. Wolf, "Design of a Real-Time Gesture Recognition System," *IEEE Signal Processing Magazine*, Vol. 22, No. 3, pp. 57–64, May 2005.doi:10.1109/MSP.2005.1425898

[112] S. Paschalakis and M. Bober, "Real-Time Face Detection and Tracking for Mobile Videoconferencing," *Journal of Real-Time Imaging*, Vol. 10, No. 2, pp. 81–94, April 2004.doi:10.1016/j.rti.2004.02.004

[113] K. Patel, "Porting PC Based Algorithms to DSPs," *Embedded Edge*, Fall 2003.

[114] K. Patel, "Porting and Optimization Techniques for C++ Based Image Processing Algorithms on TMSC62x DSP," *TI Developer Conference*, February 2004.

[115] V. Peddigari, M. Gamadia, and N. Kehtarnavaz, "Real-Time Implementation Issues in Passive Automatic Focusing for Digital Still Cameras," *Journal of Imaging Science and Technology*, Vol. 49, No. 2, pp. 114–123, March/April 2005.

[116]   V. Peddigari and N. Kehtarnavaz, "A Relational Approach to Zoom Tracking for Digital Still Cameras," *IEEE Transactions on Consumer Electronics*, Vol. 51, No. 4, pp. 1051–1059, November 2005.doi:10.1109/TCE.2005.1561824

[117]   V. Ponomaryov, F. Gallegos-Funes, O. Pogrebnyak, and L. de Rivera, "Real-Time Image Filtering with Retention of Small-Size Details and Complex Noise Mixture," *Proceedings of SPIE-IS&T Electronic Imaging Conference on Real-Time Imaging*, SPIE Vol. 4666, pp. 30–41, March 2002.

[118]   V. Ponomaryov, F. Gallegos-Funes, L. Nino-de-Rivera, and F. Gomeztagle-Sepulveda, "Real-Time Processing Scheme Based on RM Estimators," *Proceedings of SPIE-IS&T Electronic Imaging Conference on Real-Time Imaging*, SPIE Vol. 5012, pp. 37–48, April 2003.

[119]   V. Ponomaryov, A. Rosales, and F. Gallegos Funes, "Real-Time Color Imaging Using the Vectorial Order Statistics Filters," *Proceedings of SPIE-IS&T Electronic Imaging Conference on Real-Time Imaging*, SPIE Vol. 5297, pp. 35–44, May 2004.

[120]   V. Ponomaryov, R. Sansores-Pech, and F. Gallegos-Funes, "Real-Time 3D Ultrasound Imaging," *Proceedings of SPIE-IS&T Electronic Imaging Conference on Real-Time Imaging*, SPIE Vol. 5671, pp. 19–29, February 2005.

[121]   F. Porikli, "Real-Time Video Object Segmentation for MPEG-Encoded Video Sequences," *Proceedings of SPIE-IS&T Electronic Imaging Conference on Real-Time Imaging*, SPIE Vol. 5297, pp. 195–203, May 2004.

[122]   F. Porikli, "Computationally Efficient Histogram Extraction for Rectangular Image Regions," *Proceedings of SPIE-IS&T Electronic Imaging Conference on Real-Time Imaging*, SPIE Vol. 5671, pp. 36–43, February 2005.

[123]   W. Press, B. Flannery, S. Teukolsky, and W. Vetterling, *Numerical Recipes in C: The Art of Scientific Computing*. Cambridge, England: Cambridge University Press, 1992.

[124]   S. Qureshi, *Embedded Image Processing on the TMS320C6000 DSP: Examples in Code Composer Studio and MATLAB*. Berlin: Springer Verlag, 2005.

[125]   D. Rivero, M. Paindavoine, and S. Petit, "Real-Time Sub-Pixel Cross Bar Position Metrology," *Journal of Real-Time Imaging*, Vol. 8, No. 2, pp. 105–113, April 2002.doi:10.1006/rtim.2001.0259

[126]   W. Robinson and D. Wills, "Design of an Integrated Focal Plane Architecture for Efficient Image Processing," *Proceedings of the 15th International Conference on Parallel and Distributed Computing Systems, Multimedia Systems, DSP, and Image Processing Techniques*, pp. 128–135, September 2002.

[127]   R. Sangwan, R. Ludwig, P. Laplante, and C. Neill, "Performance Tuning of Imaging Applications Through Pattern-Based Code Transformation," *Proceedings of SPIE-IS&T Electronic Imaging Conference on Real-Time Imaging*, SPIE Vol. 5671, pp. 1–7, February 2005.

[128]   R. Sangwan, R. Ludwig, and C. Neill, "Software Visualization Techniques for Real-Time Imaging Applications," *Proceedings of SPIE-IS&T Electronic Imaging Conference on Real-Time Imaging*, SPIE Vol. 5671, pp. 30–35, February 2005.

[129]   O. Scherzer and A. Schoisswohl, "A Fast and Robust Algorithm for 2D/3D Panorama Ultrasound Data," *Journal of Real-Time Imaging*, Vol. 8, No. 1, pp. 53–60, February 2002.

[130]   R. Sedgewick, *Algorithms*. Reading, MA: Addison-Wesley, 1988.

[131]   J. Sivaswamy, Z. Salcic, and K. Ling, "A Real-Time Implementation of Nonlinear Unsharp Masking with FPLDs," *Journal of Real-Time Imaging*, Vol. 7, No. 2, pp. 195–202, April 2001.doi:10.1006/rtim.2000.0198

[132]   M. Smith, A. Bail, and D. Hooper, "Real-Time Image Fusion: A Vision Aid for Helicopter Pilotage," *Proceedings of SPIE-IS&T Electronic Imaging Conference on Real-Time Imaging*, SPIE Vol. 4666, pp. 83–94, March 2002.

[133]   K. Sobottka and H. Bunke, "Investigating Anytime Algorithms for Future Distance Warning Systems," *Journal of Real-Time Imaging*, Vol. 8, No. 1, pp. 61–71, February 2002.doi:10.1006/rtim.2001.0262

[134]   C. Soviany, *Embedding Data and Task Parallelism in Image Processing Applications*, Ph.D. Dissertation, Delft University of Technology, The Netherlands, 2003.

[135]   R. Stojanovic, P. Mitropulos, C. Koulamas, Y. Karayiannis, S. Koubias, and G. Papadopoulos, "Real-Time Vision-Based System for Textile Fabric Inspection," *Journal of Real-Time Imaging*, Vol. 7, No. 6, pp. 507–518, December 2001.doi:10.1006/rtim.2001.0231

[136]   T. Sumanaweera and D. Liu, "Medical Image Reconstruction with the FFT," in *GPU Gems 2: Programming Techniques for High-Performance Graphics and General-Purpose Computation*, Matt Pharr, Ed. Reading, MA: Addison-Wesley, 2005, pp. 765–784.

[137]   C. Sun, "Fast Algorithm for Local Statistics Calculation for N-Dimensional Images," *Journal of Real-Time Imaging*, Vol. 7, No. 6, pp. 519–527, December 2001.doi:10.1006/rtim.2001.0265

[138]   Y. Sun, H. Zhang, and G. Hu, "Real-Time Implementation of a New Low-Memory SPIHT Image Coding Algorithm Using DSP Chip," *IEEE Transactions on Image Processing*, Vol. 11, No. 9, pp. 1112–1116, September 2002.doi:10.1109/TIP.2002.802533

[139] P. Tatzer, T. Panner, M. Wolf, and G. Traxler, "Inline Sorting with Hyperspectral Imaging in an Industrial Environment," *Proceedings of SPIE-IS&T Electronic Imaging Conference on Real-Time Imaging*, SPIE Vol. 5671, pp. 162–173, February 2005.

[140] C. Torres-Huitzil and M. Arias-Estrada, "Reconfigurable Vision System for Real-Time Applications," *Proceedings of SPIE-IS&T Electronic Imaging Conference on Real-Time Imaging*, SPIE Vol. 4666, pp. 124–132, March 2002.

[141] P. Tsai, C. Chang, and Y. Hu, "An Adaptive Two-Stage Edge Detection Scheme for Digital Color Images," *Journal of Real-Time Imaging*, Vol. 8, No. 4, pp. 329–343, August 2002.doi:10.1006/rtim.2001.0286

[142] L. Tsap, "Gesture-Tracking in Real Time with Dynamic Regional Range Computation," *Journal of Real-Time Imaging*, Vol. 8, No. 2, pp. 115–126, April 2002.doi:10.1006/rtim.2001.0260

[143] I. Urriza, J. Artigas, L. Barragan, J. Garcia, and D. Navarro, "VLSI Implementation of Discrete Wavelet Transform for Lossless Compression of Medical Images," *Journal of Real-Time Imaging*, Vol. 7, No. 2, pp. 203–217, April 2001.doi:10.1006/rtim.1999.0171

[144] I. Uzun and A. Amira, "Real-Time 2-D Wavelet Transform Implementation for HDTV Compression," *Journal of Real-Time Imaging*, Vol. 11, No. 2, pp. 151–165, April 2005.doi:10.1016/j.rti.2005.01.001

[145] S. Venugopal, C. Castro-Pareja, and O. Dandekar, "An FPGA-Based 3D Image Processor with Median and Convolution Filters for Real-Time Applications," *Proceedings of SPIE-IS&T Electronic Imaging Conference on Real-Time Imaging*, SPIE Vol. 5671, pp. 174–182, February 2005.

[146] M. Wei and A. Bigdeli, "Implementation of a Real-Time Automated Face Recognition System for Portable Devices," *Proceedings of the IEEE International Symposium on Communications and Information Technologies*, pp. 89–92, October 2004.

[147] P. Whelan, P. Soille, and A. Drimbarean, "Real-Time Registration of Paper Watermarks," *Journal of Real-Time Imaging*, Vol. 7, No. 4, pp. 367–380, August 2001.doi:10.1006/rtim.2000.0239

[148] J. Wakerly, *Digital Design: Principals and Practices*. Englewood Cliffs, NJ: Prentice Hall, 2000.

[149] J. Wickramasuriya, M. Alhazzazi, M. Datt, S. Mehrotra, and N. Venkatasubramanian, "Privacy-Protecting Video Surveillance," *Proceedings of SPIE-IS&T Electronic Imaging Conference on Real-Time Imaging*, SPIE Vol. 5671, pp. 64–75, February 2005.

[150] P. Wong and M. Bhavana, "Hardware in Process: Mobile Handset Cameras Challenge Image Processors," *Optical Engineering Magazine*, Vol. 5, No. 9, pp. 15–17, October 2005.

[151] Y. Xu, Y. Jia, and W. Liu, "Person Real-Time Tracking for Video Communication," *Proceedings of SPIE-IS&T Electronic Imaging Conference on Real-Time Imaging*, SPIE Vol. 4303, pp. 35–42, April 2001.

[152] M. Yang, T. Gandhi, R. Kasturi, L. Coraor, O. Camps, and J. McCandless, "Real-Time Implementation of Obstacle Detection Algorithm on a Datacube MaxPCI Architecture," *Journal of Real-Time Imaging*, Vol. 8, No. 2, pp. 157–172, April 2002.doi:10.1006/rtim.2001.0272

[153] R. Yang and M. Pollefeys, "A Versatile Stereo Implementation on Commodity Graphics Hardware," *Journal of Real-Time Imaging*, Vol. 11, No. 1, pp. 7–18, February 2005.doi:10.1016/j.rti.2005.04.002

[154] J. Zhang and W. Liu, "Real-Time Image Sequence Segmentation Using Curve Evolution," *Proceedings of SPIE-IS&T Electronic Imaging Conference on Real-Time Imaging*, SPIE Vol. 4303, pp. 67–78, April 2001.

[155] F. Ziliani and A. Cavallaro, "Image Analysis for Video Surveillance Based on Spatial Regularization of a Statistical Model-Based Change Detection," *Journal of Real-Time Imaging*, Vol. 7, No. 5, pp. 389–399, October 2001.doi:10.1006/rtim.2000.0208

[156] AccelChip, Inc., *MATLAB DSP Algorithmic Synthesis for FPGAs and ASICs*, http://www.accelchip.com/, 2005.

[157] Catalytic, Inc., *Catalytic MCS Family: MATLAB to C Code Synthesis*, http://www.catalyticinc.com/, 2005.

[158] GPGPU, *General-Purpose Computation Using Graphics Hardware*, http://www.gpgpu.org/, 2005.

[159] Texas Instruments, Inc., *TMS320DM320 CPU and Peripherals Technical Reference Manual*, 2004.

[160] Texas Instruments, Inc., *DaVinci Technology*, http://www.ti.com/davinci/, 2005.

About the Authors

Nasser Kehtarnavaz is a Professor of Electrical Engineering at the University of Texas at Dallas. He has written four other books and numerous papers pertaining to signal and image processing, and regularly teaches undergraduate and graduate signal and image processing courses. Among his many professional activities, he is currently serving as Coeditor-in-Chief of *Journal of Real-Time Image Processing*, and Chair of the Dallas Chapter of the IEEE Signal Processing Society. Dr. Kehtarnavaz is a Fellow of SPIE, a Senior Member of IEEE, and a Professional Engineer.

Mark Gamadia is currently a Ph.D. Candidate and an Erik Jonsson Research Assistant Fellow in the Department of Electrical Engineering at the University of Texas at Dallas. He has written several papers involving the development and real-time implementation of image processing algorithms. Mr. Gamadia is a Student Member of IEEE, IEEE Signal Processing Society, and IEEE Consumer Electronics Society.

Printed in the United States
by Baker & Taylor Publisher Services